THE FORGOTTEN CAVES

OF

BISBEE, ARIZONA

A REVIEW OF THE HISTORY AND

THE GENESIS OF THESE UNIQUE FEATURES

Richard W. Graeme III

Douglas L. Graeme

Richard W. Graeme IV

First Edition July 6, 2009
Second Edition April 9, 2016
Copper Czar Publishing

COVER PHOTOGRAPHS:

OXIDATION CAVE - 7TH LEVEL, SOUTHWEST MINE, TAKEN BY PETER L. KRESAN AND RICHARD W. GRAEME III IN 1977.

TABLE OF CONTENTS

FOREWORD

Bisbee's indisputable position as America's best and most important source of fantastic minerals was established, in no small part, by the specimens recovered from the many caves found during the first 50 years of mining. Yet very little has been written about these unusual caves, which were formed by the oxidation of the sulfide mineral deposits. Indeed, these uniquely beautiful caves are largely forgotten. Because of the continuing movement of the overlying rock during mining and the need to backfill all openings for safety reasons; none of Bisbee's many caves remain accessible today. The intent of this book is to bring together as much of the history and as many of the firsthand accounts as possible, to record the grandeur of these extraordinary caves. At the same time, an effort is made to explain the unusual oxidation origin of the caves in Bisbee and to describe their distinctly uncommon beauty.

A sense of the greatness of Bisbee's oxidation caves can be had by looking at the specimens recovered prior to mining, notably those incorporated in oxidation cave reproductions. Today, a very fine recreation of a typical, small, oxidation cave from Bisbee is on display in the Janet Hooker Mineral Hall of the U.S. National Museum of Natural History (Smithsonian) in Washington, D.C. A second, and equally fine, oxidation cave display can be seen in the Bisbee Mining and Historical Museum in Bisbee. Both displays were developed using a number of fine specimens recovered from some of these caves, often more than a century ago. All of the authors served as consultants to both museums during the design and construction of these important exhibits.

Our interest in the oxidation caves in Bisbee began long before being asked to assist in the design of the two displays noted above. All of the authors were born in Bisbee, Richard III in 1941, and Richard IV and Douglas, often referred to as the twins, in 1968. The family ties to this

Figure F-1: Recreated oxidation cave in the Bisbee Mining and Historical Museum. Courtesy of Bisbee Mining and Historical Museum photo by John Harris.

fascinating community extending back to 1883 when Charles Keeler, Richard III's maternal grandfather, first came to here to work in the mines. Charles collected minerals from the mines, but most importantly, his mining stories were inspirational, if not catalytic to the process of the family's continuing passion for Bisbee, its history, minerals and unique geologic features.

Richard Graeme III began exploring the Bisbee mines and collecting minerals in 1948, at a very young age and visited many oxidation caves over the years. Then, in 1960, Richard III began working underground as a miner in Bisbee, which he did for a dozen years before becoming the resident geologist at the Copper Queen Branch in 1972. This came with a BS degree in geological engineering earned at the University of Arizona in Tucson in 1972, all the while, working full time as Bisbee miner.

In 1974, a mineral species new to science from Bisbee was named "graemite" for Richard III to acknowledge both its discovery and his contributions made to the science of mineralogy. Since that time, several important articles on Bisbee minerals (Graeme 1981 and 1993) were published, both with the assistance of the other authors, Douglas Graeme and Richard Graeme IV. In 2015, all three of the authors collaborate in writing yet an additional update on the mineralogy of Bisbee (Graeme, Graeme & Graeme, 2015).

Douglas and Richard IV began to venture underground in Bisbee at the age of six in 1974. The first place they visited contained oxidation caves. Soon thereafter, we all began the exploration of the caves associated with the New Southwest Orebody, which are discussed in detail in Chapter Seven of part one. Over the next 30 years, we made innumerable trips into these mines observing the mineralogy and studying the geology.

Both Douglas and Richard IV worked as underground miners in Colorado and New Mexico as well as in the gold fields of Nome, Alaska while pursuing their education. Today, Douglas is the manager of the Queen Mine Tours in Bisbee while Richard IV teaches science in the nearby Sierra Vista schools.

Over the last 60 plus years, the authors have cooperatively assembled the most comprehensive scientific collection of Bisbee ores and minerals in existence (Moore, 2006). An important segment of this research collection is composed of minerals recovered from many of the oxidation caves. Most of these specimens were salvaged from the caves, just ahead of mining, from the early 1880s through the first three decades of the 20th century. The importance of this research base to this and similar works cannot be overstated as it provides a ready and well-documented source of study material.

ACKNOWLEDGMENTS

We began our work on this book by carefully researching the abundant literature on this rich mining district, including the files of several organizations. Relevant files held by Phelps Dodge Corporation (PD) at the Copper Queen Branch were studied while Richard Graeme III was employed by PD, as a geologist. Also, materials held by the Bisbee Mining and Historical Museum, as well as the Arizona Historical Society in Tucson were reviewed. Collections of historic images in the Bisbee Mining and Historical Museum, the Bisbee Restoration Museum, Arizona Historical Society, American Museum of Natural History, Jeremy Rowe and Bob Jones were reviewed in an effort to better understand the many caves they represented. Access to these collections by the institutions and private collectors is gratefully acknowledged.

Because specimens from these caves are disseminated worldwide over the last century, the voluminous Bisbee minerals in most of the important museums and many private collections, throughout the world, were studied in the context of their relationship to the caves and associated ore deposits. The authors are deeply indebted to the very cooperative staff in charge of the minerals at the U.S. National Museum, Natural History (Smithsonian), American Museum of Natural History, Los Angeles County Museum of Natural History, British Museum of Natural History, Harvard Mineral Museum, Arizona Sonora Desert Museum, and University of Arizona Mineral Museum. Also, many individual collectors, too numerous to name, are thanked for their many kindnesses in providing access to their fantastic specimens for study purposes.

Peter Kresan, retired Senior Lecturer in the Department of Geosciences at the University of Arizona did the initial review of our book. Peter is a longtime friend who has spent many hours underground in Bisbee with us. Several of Peter's extraordinary photographs, including the cover photograph, all taken more than 30 years ago, are included. A very special thanks to Pete for these important contributions.

Unless otherwise credited, the historical photographs in this book are from the Graeme-Larkin Collection, while the mineral specimens illustrated are from the Graeme Collection, and the mineral photographs are by the authors.

Cave mineral specimens are presented in their historical context, as artifacts, reflecting the foresighted recovery efforts made by mining companies, often in cooperation with museums during the early years of mining. The authors are decidedly of the opinion that, except for such rare circumstances, as when destruction is inevitable, the removal of minerals from caves is inappropriate and must not be condoned. Nothing presented within this book is to be considered contrary to this anti-removal position, which we have long held.

INTRODUCTION

Hundreds of caves were found while mining the rich oxide copper ores in Bisbee. Often these openings were beautiful beyond description, as they were typically well decorated, but there was more. In many of the caves, some of the speleothems, such as stalactites, stalagmites and other cave formations, were stunningly tinted in varying hues of blues and greens by the ever-present copper. On rare occasion, nearly the whole cave contained formations so colored (Beasley, 1916). This remarkable coloring was invariably complemented by a striking red-brown hue in some speleothems because of iron oxide inclusion, creating a handsome contrast with the other, mostly white formations.

Figure I-1: Calcite stalactites tinted green by copper. Minor malachite is also on and included within the calcite, 5th level, Southwest Mine, view – 14 inches.

In part one of this book, firsthand accounts of these incredible caves have been collected and put in their historical context. In this context, we discuss the fascination with these caves at the time of their discovery and the mystery which originally surrounded their origin as well as that of the rich copper ores found beneath the caves. And too, the reason for the necessary destruction of

6

these caves will be put in the perspective and the context of the time in an effort to separate the sad reality of destruction from the impractical ideal of preservation.

As will be detailed in the second part of this book, the authors are convinced that the origin of these caves is absolutely a function of the complete, near-surface oxidation of the sulfide replacement deposits in the several limestone formations. Because of this, oxidation caves occurred above large bodies of thoroughly oxidized copper ore. Caves were also found above huge masses of iron oxides, with little if any copper ore present. This is a reflection of the original, largely pyrite, mineralogy of the sulfide masses prior to oxidation.

Economic geologist and miners, like ourselves, have long held the opinion that the caves in Bisbee formed in this manner. In 1927, the economic geologist, Edward Wisser, went to great lengths to convincingly demonstrate what others had long known regarding the formation of these caves. However, this mechanism for cave development remains a controversial topic in the science of cave formation (speleogenesis) with many well-respected cave scientists skeptical of the position.

Several illustrations after Wisser (1927) are presented below to frame our position. First is a comparison between the typical way a normal cave develops by surface water solutions and one developed through sulfide oxidation (Figure I-2). Secondly, is an idealized cross section of an oxidation cave with many of the more important features related to the development illustrated (Figure I-3). Above all, note the position of the oxidation cave over the oxidized orebody. The fracturing of overlying limestone with marginal cracks and the separation and opening up of the rock layers or beds, along doming cracks. All of these are discussed in detail later.

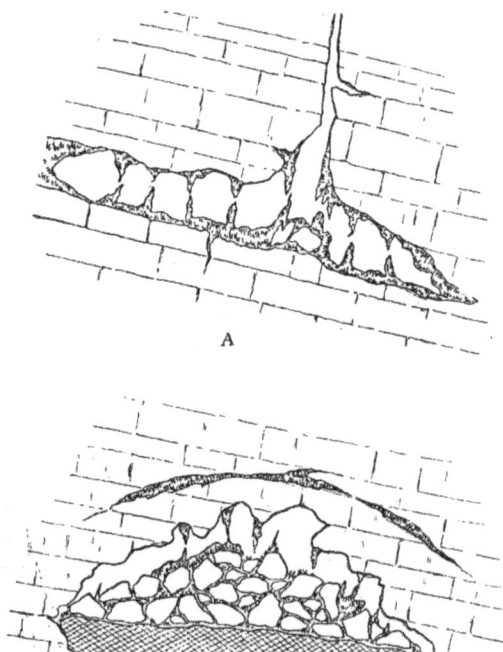

Figure I-2: Comparison of a normal solution developed cave (A) with an oxidation cave (B). After Wisser (1927).

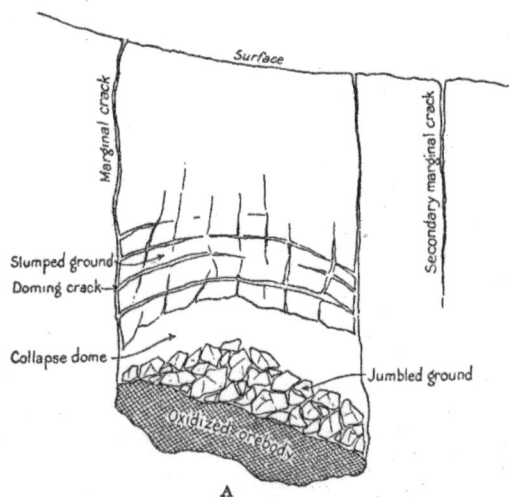

Figure I-3: Features typically found with an oxidation cave development, notably the oxidized orebody below the cave, marginal cracks, slumped ground, doming cracks and jumbled ground. After Wisser (1927).

The total number of oxidation caves discovered during mining is not known. The authors estimate that hundreds and perhaps in excess of a thousand oxidation caves were found in Bisbee between the start of mining in 1880 and exhaustion of completely oxidized ores in about 1930.

Even though these caves were common, their splendor was always appreciated by all who saw them; miners and visitors alike. Occasionally, the splendor of a cave was recorded by written descriptions or photographs. Herein are a few of the accounts, as written by those who visited them as well as period photographs of several of the caves described and others. As was often noted, it was impossible to adequately describe the wonder of these caves with words, but they attempted to do so, as best they could. These stories are shared later in this book.

Colorful stalactites and other cave formations were collected, in truth, salvaged, by the mining companies, the miners themselves, museums and mineral dealers. Under most circumstances, it would have been completely inappropriate to remove formations from these caves, much less completely ruin them. Indeed, it would have been nothing short of vandalism, as recorded by some tasked with this salvage effort. However, for a number of safety and economic reasons, these caves were destroyed. Essentially the caves were destroyed because the great majority either contained or occurred over substantial amounts of ore. Had the ore not been mined and left in place to preserve even one cave, the costs in both economic and social terms would have been enormous, indeed, unacceptably high. And too, this would have created a potential safety hazard.

The obvious loss of revenue from the large amounts of ore found, yet not mined to preserve a single cave would have placed the mine in an economically precarious position. Men may have lost their jobs. This would have had huge economic repercussions well beyond Bisbee. The Copper Queen Mine was the economic engine of southern Arizona at the time.

Also, the high costs associated with maintaining access to the cave would have been unsustainable. Maintaining safe access through the mine shafts and other openings (crosscuts) in these soft, ore areas where the caves occurred, would have been both difficult and inordinately expensive. Huge amounts of support timber and constant maintenance would have been required. For example, 30 board feet of support timber was required for every ton of ore removed from the mine and this was just to keep the mine workings open on a short-term basis while the ore was being extracted (Bisbee Daily Review, 1904).

During mining operations, oxidation caves actually presented a significant hazard. In unstable rock conditions, such as those found at Bisbee, all openings of any size must be backfilled (tightly filled with mined waste rock) to prevent the movement and collapse of ground (rock). Such movement could dangerously affect nearby areas. Also, as these caves formed, natural subsidence fractured the rock hosting the cave for up to hundreds of feet above, making the area surrounding the cave relatively weak and largely unstable. This pervasive fracturing is clearly illustrated by Figure I-3

(Wisser, 1927). He further notes that the effects of subsidence extended as much as 1000 feet above the oxide orebodies and associated cave.

For these reasons, the cave areas were mined and the remaining openings backfilled. Today, the only records that remain of these caves include the awe-inspired descriptions written by those fortunate enough to see them; historic photographs, mostly in black and white, and the many specimens recovered before the caves were lost during mining.

Figure I-4: Stalactites and helictites tinted by copper and iron. Calcite is colored green by copper or red-brown by iron, while aragonite is tinted blue-green by copper, 100 level, Holbrook Mine, view – 15 feet.

This discussion will start with a chronicle of the stories about the caves from the time they were first discovered. Our historical review will include descriptions by those who saw them and corresponding photography, when available. Other period references from mining trade publications and mineral collecting journals of the time are also included. The historical information is presented in chronological order, by decade, from the 1880s to the 1930s.

By the early 1930s, very few caves were being found due to the fact that mining had passed through the zone where oxidation was complete enough to allow for cave development. Nevertheless, the authors were fortunate enough to visit a number of the remaining caves over a period of nearly 60 years, a few of which will be dis as to their characteristics and their individual features.

The second part of this book is devoted to a more technical discussion on the formation of oxidation caves and a review of the minerals found associated with the caves. The intent here is to set out the argument, with supporting evidence, that the Bisbee caves were indeed formed as the result of the complete oxidation of sulfide replacement deposits and not preexisting caves into which later ores were deposited, as some believe.

PART 1: A HISTORICAL OVERVIEW OF THE OXIDATION CAVES IN BISBEE, ARIZONA.

Chapter One: The 1880s

The serendipitous discovery of mineralization, in what would become Bisbee, in the summer of 1877 did little to foretell of the great beauty just below the surface of this remote, desert canyon. The first hint came in 1880 with the accidental discovery of extraordinarily rich copper ores on the Copper Queen Mining Claim, were brush was being removed (Douglas, 1881). Mining began, in earnest, soon thereafter.

Rich mineral deposits in limestone such, as found on the Copper Queen claim were rare and, at the time, the geology was poorly understood. No one could conceive of a way for such rich ores to have formed in otherwise seemingly barren limestone. Ideas regarding the replacement of limestone by mineralizing fluids, such as happened at Bisbee, were just evolving and the concept of the oxidation

Figure 1.1.1: Working in the open cut of the Copper Queen Mine, 1881.

of sulfides and related processes was far from universally understood or accepted. The early views of how the Copper Queen Orebody formed were, founded in geologic concepts of very different deposit types. Dr. James Douglas, a man of some note in the copper business of the era, wrote the first account of these ores and of a cave.

Dr. Douglas first came to the remote mining camp of Bisbee in early February 1881. He arrived soon after serious mining began to examine the Copper Queen Mine for a potential investor. Douglas could not explain what he saw geologically. By now, the mine opening was cave-like with smooth walls where the ore had once been held (see Figure 1.1.1) and this undoubtedly influenced his thinking. While attempting to describe the geology, Dr. Douglas wrote:

> *"In the limestone also many bootless* [fruitless] *attempts have been made to open mines on carbonate* [malachite, a copper carbonate] *croppings, which, promised fair; but led only to small deposits, filling apparently such cavities as are common in all limestone formations.*
>
> *Copper glance* [chalcocite, a copper sulfide] *is also in places sparingly scattered, thro the limestone, but has nowhere 'been found in large or regular' bodies. I think*

that it was a cave, only a very large one, which received, this mineral, that now
makes the Copper Queen a mine, and that its occurrence as a carbonate, is due to
the limestone, which originally carried it and from which it filtered into this
reservoir. This conjecture be correct, there is no reason to suppose that sulphurets
[sulfides] *will be found."* (Douglas, 1881)

No doubt, the presence of caves in association with the Copper Queen Orebody helped form Douglas' opinion of the genesis of this deposit. He noted a cave in the Copper Queen Mine by reporting:

"At 40 feet from the mouth of this drift, a stalactite cavern was struck. It is about
18" wide. The same cavern has been struck, as I shall show, at 59 feet below the
level of this drift, and I was able to scramble up it for about 20 feet. This south drift
follows it for 42 feet. At the point where this cave was struck levels were started to
east and west. That to the West [sic] *follows the cave for 31 feet in a Northeast and*
Southwest direction, and is throughout in good ore. The ore in contact with the
stalactite casing of the cave is partly cuprite. The level to the east runs for 54 feet
in good ore" (Douglas, 1881).

Dr. Douglas' interpretation of the origin of these ores was incorrect, but then he had so little to guide him. He saw the Copper Queen deposit in the context of limestone cave, hosted ore deposits being mined in the mid-west at the time. Within a few years and with a good deal more information, Dr. Douglas correctly interpreted both the origin of the ores and of the caves so often associated with the oxide ores.

Over the next several years, few accounts were recorded of the caves found in Bisbee. Those reported were but brief notes, generally in efforts to publicize new ore finds. The hope was to attract the interest of potential investors in the property, as these small mines were always in need of additional capital. Most of the notes were printed in *The Engineering and Mining Journal* (E&MJ). It was the most widely circulated and most reputable mining trade magazine then, as it is today. In the March 26, 1881 issue of this prestigious magazine, it was reported that in the Copper Queen Mine:

"At 40 feet, a cave was struck branching east and west. They have drifted from this
point both ways, 63 feet west and about 60 feet east, both in good ore."

It is almost certain that this cave is the same one Douglas visited, but after additional mining had been carried out by extending the workings in both directions.

In January 1883 *The Engineering and Mining Journal* noted the occurrence of substantial ore in the "Cave" Mine, later renamed the Holbrook, with the following:

"From Bisbee, it is reported that in this mine [Cave], on the eastern slope of Copper Queen Hill, a large cave containing considerable ore has been discovered."

In 1886, the noted geologist Arthur Wendt visited Bisbee, as well as other southwestern copper deposits. He, like James Douglas, became convinced that the ores had been deposited in pre-existing caves because of the close association of the ore with these caves (Wendt, 1887). Wendt had much more to see than Douglas saw five years earlier as more caves had been found and the first sulfide ore discovered, but these did not seem to have influence his view. Wendt did describe two Bisbee caves, including what he considered an unusual one as it was not associated with ore. He wrote:

"In other places, what have evidently been vugs and caves in the bedded limestone have been filled by the ore. Not all caves near the ore bodies contain ore. A notable one, east of and near the big Copper Queen ore-body, has perfectly smooth and rounded water-worn sides; and in the bottom of this cavity, round boulders, gravel, and sand of limestone and quartz are found, as if this cavity had formally been the seat of intense geyser action and the materials now lying in the bottom of the cave were formally violently churned in it."

Figure 1.1.2: Wall of the cave on the "A" level of the Copper Queen Mine described by Wendt, (1887). Note the rounded, clearly solution-shaped features as discussed by Wendt. Some minor, local botryoidal calcite coating is evident as well, vertical view – 11 feet.

The cave with no ore described above by Wendt is located on the "A" level of the Copper Queen Mine and has been visited several times by the authors. It is closely associated with barren iron oxides and manifests the characteristics common to oxidation caves. The rounded boulders, pebbles and sand noted by Wendt (1887) are not common in oxidation caves, but some do contain these materials such as that shown in Figure 1.1.3. These are in a small cave in the nearby Copper Prince Mine. Boulders, pebbles and sand were also noted in oxidation caves by Douglas (1900) when he wrote:

"...there is abundant evidence that, through cracks and fissures in the limestone, surface-detritus has been carried to considerable depth."

The cracks and fissures referred to by Douglas above were the marginal cracks associated with the oxidation subsidence as explained by Wisser (1927) and as shown in Figure I-3. Wisser, (1927) also noted that sand, gravel and conglomeritic material was common in the caves.

Figure 1.1.3 Stratified sand, pebbles and boulders, deposited by surface inflow, in the bottom of a small cave, Copper Prince Mine. A wide range of rock types is present in this detritus, horizontal view – 15 feet.

Also, smooth walls such as found in the cave on "A" level are a very common feature in many oxidation caves and not necessarily the product of abrasion from running water.

Wendt continued with a hypothetical description of the Copper Queen Orebody being deposited in a pre-existing cave, as he supposed had happened. Examples of the ores from this orebody are illustrated in Figures 1.1.4 and 1.1.5 to show, at least in part, how he might have arrived at his conclusion.

In Figure 1.1.4, the top, left surface of the malachite was in contact with the smooth limestone wall of the Open Cut with but a thin layer of yellowish clay between the limestone and the malachite. This was typical for this orebody at the ore/limestone boundary. The plumose growth patterns of the malachite indicate several periods of in situ deposition, starting on the limestone wall rock, as the apparently high copper sulfides oxidized. Calcite was deposited last, lining most open spaces in the malachite and covering the outer most surface.

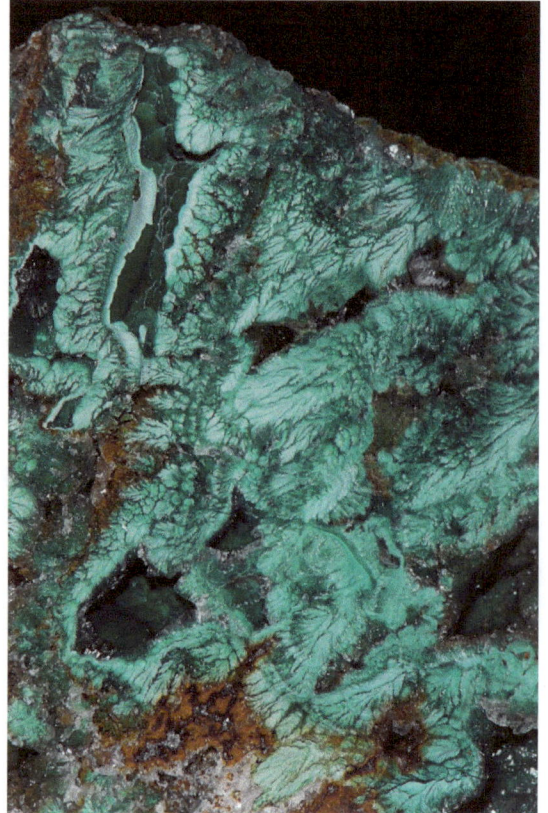

Figure 1.1.4: Example of malachite ore with later calcite from of the wall of the Open Cut of Copper Queen Mine, of the type described by Wendt, (1887), vertical view – 4 inches.

"To those who have studied the deposition of ores in the district, there can be no doubt that there is some relation between the existence of these cavities in the limestone and the deposition of the ore. When the Copper Queen was first opened, a horizontal drift and open cut was run in on the bedding of the limestone, following the outcrop of the copper ore. Only a little while in, the ore began to rise, and the work eventually formed what is now the great open cut of the Copper Queen Mine.

Formerly, no doubt, a huge cave occupied this place the bottom and sides of which were gradually covered with ore. From the roof of this cave, stalactites of carbonate of lime were dependent; and in parts of the cave these stalactites contained not only carbonate of lime but carbonate of copper as well, the ore occurring in the shape of a series of concentric rings, the carbonate of lime always finally covering the copper ore."

Figure 1.1.5 Stalactites of malachite covered by calcite in the manner described by Wendt, (1887) from the Copper Queen Mine Open-Cut, view - 1.7 inches. Gene Wright collection.

15

Fortunately, numerous specimens are available for study, as the discovery and the mining of copper deposits at Bisbee occurred during a time of a great curiosity about all things natural. It was an age of scientific discovery and interest in the minerals and the caves, which held them, was universally high. There was a desire by many to collect these lovely, natural curiosities, and thus a market developed for mineral specimens. This led to the recovery and preservation of many specimens from the caves before their inevitable destruction. It was also most fortuitous that Dr. James Douglas was in charge of what was now the Copper Queen Consolidated Mining Company (Copper Queen), because he had an appreciation for the importance of the exceptional minerals found at Bisbee. Douglas saw to the recovery of many thousands of specimens from Bisbee's caves by using company employees to collect them for distribution to institutions worldwide. Importantly, he also encouraged commercial collecting of the caves. However and perhaps most significantly, the Copper Queen always allowed the miners to collect these minerals as well, so long as they did so without neglecting their duties or taking undue risks. Bisbee's miners collected thousands and thousands of mineral specimens from the many mines in the district during the near century of operation because of this enlightened policy.

Perhaps the earliest report of cave specimens from Bisbee being appreciated as collectable mineral specimens is recorded in the *Transactions of the New York Academy of Science* in 1889. The minutes of the regular business meeting for January 7, 1889 note that:

> *"Dr. James J. Friedrich exhibited and described some remarkable minerals from the copper district of Arizona, as follows:*
>
> *I wish to present a few specimens from Bisbee, Arizona, viz.*
>
> *(1) a beautiful green calcite and (2) an aragonite stalactite: The calcite is imbedded in a mass of malachite; in the centre of the stalactitic columns, fine apertures are observed, around which silky fibers of malachite are arranged concentrically Through these apertures malachite penetrated the columns, coloring them in some instances only in the centre, while in other columns the whole mass of the calcite is penetrated and tinted. The most beautiful coralloidal specimens of aragonite come from the Silver Spray Mine, which is at present not accessible; private collections in Tombstone contain very fine large specimens, resembling a good sized coral stem, with its branches perfectly white and translucent. In this region of the carbonates, It is very natural that pseudomorphs and imitative shapes should be very frequent; and the one mineral which is best adapted for the purpose is azurite. Pure massive azurite, translucent and bearing in its crystallized form a close resemblance to epidote, is rarely found. This specimen (3) in point I obtained at the Silver Bear Mine, at which mine exclusively chalcocite also occurs.*

(4.) The next specimen is a perfect pseudomorph of azurite after calcite; I procured it from a dealer in San Francisco, who would not tell me the locality where it was found, but I presume it to be from Black Co., N. M., where fine specimens of azurite are not rare.

(5.) These specimens are oolitic in structure; the material generally found to build up the different layers of the concretions are tenorite, malachite and azurite. The same construction is seen in No.6 in stalactitic shapes and incrustations. No.7 is one of the most instructive specimens exhibiting all the various degrees of metamorphism. This piece of ore shows in its totality the structure of fibrous aragonite the principal mass is penetrated by azurite, but at the same time one may observe nodules of tenorite; patches of malachite, in some instances coarsely fibrous, and in one place a bluish-green concentration, which forms a new mineral, the so-called calco-malachite, of which you have here (8) a larger and handsomer specimen. No. 7, above described, is therefore a truly representative specimen of the Bisbee copper region.

These are not by any means the handsomest specimens obtained from Bisbee; fine minerals from this locality are by no means rare; still, I succeeded in getting some choice material from old excavations, such as a very beautiful malachite stalactites, etc., which may compete with any Bisbee specimens ever exhibited.

The cave specimens described above seem almost too extraordinary to comprehend. They were however undoubtedly very similar to the calcite coated, malachite stalactites illustrated in Figure 1.1.5. It is surprising that the Silver Spray and Silver Bear mines are noted as sources of good specimens. By late 1888, the Silver Spray had operated only briefly and was closed because of a lack of ore, while the Silver Bear was almost totally barren of ore. A few years later, the Copper Queen purchased both the Silver Spray and Silver Bear mines. With more exploration work, the Silver Spray became a very important mine, but the Silver Bear remained a disappointment.

Friedrich's comment about specimens from Bisbee being in private collections in Tombstone, notably cave type coralloidal aragonite, is interesting. At the time these specimens were acquired by Dr. Friedrich (1888), Tombstone was a larger and far more important mining community than Bisbee. No doubt, many of its residents appreciated and collected minerals. Nearby Bisbee would have been a convenient source. Also, it is possible that the stalactitic malachite specimen mentioned by Friedrich also came from a Bisbee oxidation cave. This was a common occurrence for malachite at Bisbee during the early years.

Chapter Two: The 1890s

Bisbee became justifiably famous for its fine and strikingly beautiful minerals soon after mining began. It produced some of the best examples of malachite and azurite, found anywhere. The quality of the azurite, malachite, chrysocolla and cuprite found here is described in Figure 1.2.1, which are several excerpts from the George English *Catalogue of Minerals for Sale*, from 1890.

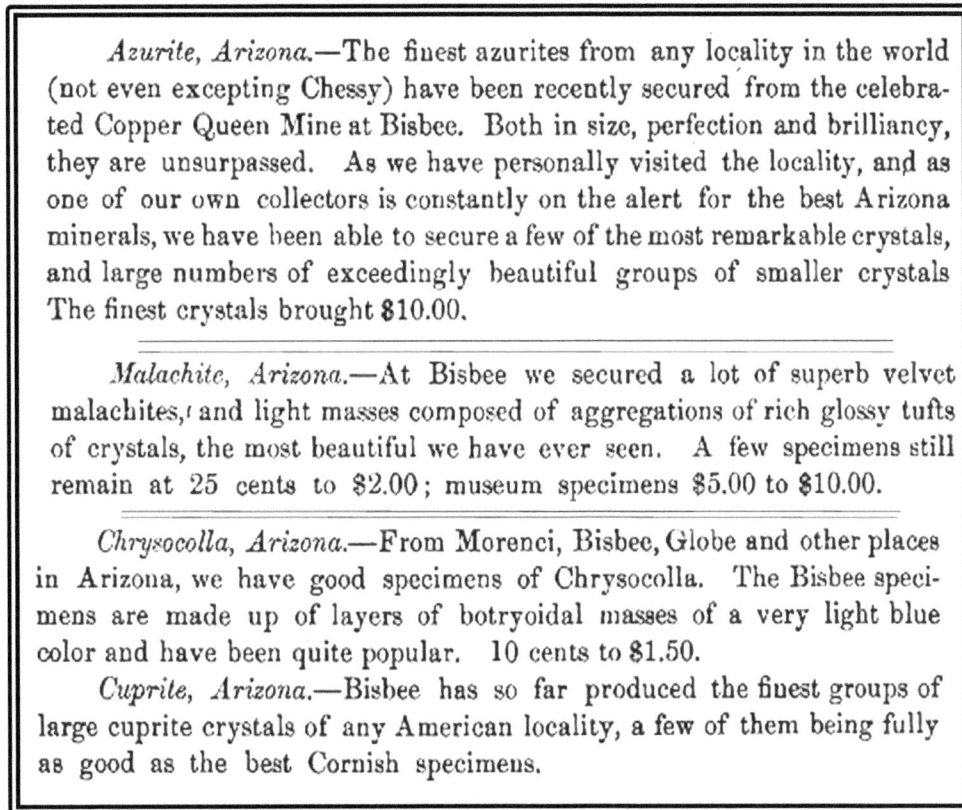

> *Azurite, Arizona.*—The finest azurites from any locality in the world (not even excepting Chessy) have been recently secured from the celebrated Copper Queen Mine at Bisbee. Both in size, perfection and brilliancy, they are unsurpassed. As we have personally visited the locality, and as one of our own collectors is constantly on the alert for the best Arizona minerals, we have been able to secure a few of the most remarkable crystals, and large numbers of exceedingly beautiful groups of smaller crystals The finest crystals brought $10.00.
>
> *Malachite, Arizona.*—At Bisbee we secured a lot of superb velvet malachites, and light masses composed of aggregations of rich glossy tufts of crystals, the most beautiful we have ever seen. A few specimens still remain at 25 cents to $2.00; museum specimens $5.00 to $10.00.
>
> *Chrysocolla, Arizona.*—From Morenci, Bisbee, Globe and other places in Arizona, we have good specimens of Chrysocolla. The Bisbee specimens are made up of layers of botryoidal masses of a very light blue color and have been quite popular. 10 cents to $1.50.
>
> *Cuprite, Arizona.*—Bisbee has so far produced the finest groups of large cuprite crystals of any American locality, a few of them being fully as good as the best Cornish specimens.

Figure 1.2.1: A collage of excerpts from *Catalogue of Minerals for Sale*, (English, 1890).

Mineral specimens of all types, including cave specimens, from Bisbee were in great demand by museums and collectors alike. The U. S. National Museum Report for the year ending June 30, 1890 notes the museum received:

> *"Two hundred and sixty specimens of malachite, azurite, cuprite, calcite crystals and stalactite from the Copper Queen Mine, at Bisbee, Arizona. (22897). Collected by Mr. W. P. Jenny."* (Smithsonian, 1891)

Many collectors and other museums wanted to acquire similar pieces for their own collections. Thus, mineral dealers were eager to get whatever they could and many made the difficult trip to this remote settlement in the distant Arizona Territory in search of such material. Some dealers sent field agents on their behalf, to seek out and purchase the minerals for them to sell. Most prominent among these was the noted New York mineral dealer George English who sent William Niven to Bisbee in late 1890.

William Niven collected and purchased numerous mineral specimens for Mr. English, but it was the cave minerals he collected that gained the most attention. Niven recovered specimens from a cave in the Czar Mine, beneath Queen Hill. In doing so, he gave us the earliest detailed description of one of Bisbee's oxidation caves. An excerpt from a letter he had sent to English was published in the July 1891 issue of *Mineralogist's Monthly* and offered the following:

> *"It is 80 feet wide at the mouth, 270 feet wide about the center and 250 feet wide at the end, and is 500 feet in length. From the mouth of the cave to the end it is at least 150 feet high. There are four chambers, each of which has its own peculiar habit of crystal form. In No. 1 are to be found the acicular crystals of aragonite. No. 2 consists mainly of flos ferri. No. 3 is the grandest of all, and looks like a magnificent cathedral-most of the stalactites and stalagmites are colored green with the copper, and they look like immense organs, while hanging from the roof is a bunch of stalactites which looks like a gigantic chandelier. In some places the form is like roses, again like fringe, coral, palm leaves, trees, toadstools; in others great slopes of glaciers and fields of ice. . . . In one place more is a great number of sheets of aragonite from 3 to 4 feet long and 2 to 3 feet wide, about 1/2 to 3/4 inch thick, beautifully translucent, showing alternate layers of green, white and blue and resembling tapestry when a light is placed behind it. . . . I spent 14 days in this vast cavern. While I was selecting specimens I had a photographer take pictures; you can imagine the singular scene the main chamber presented with 200 candles burning. It was a scene of dazzling beauty. Countless stalactites of every conceivable shade of green, blue, red-intermingled by snow-white, hanging from the roof and sides-while rising from the ground great ghost-like stalagmites stood silent sentinels guarding this incomparable workshop of nature's laboratory."*

Figure 1.2.2: William Niven (left with tie) in a cave in the Czar Mine, 1890.

Unfortunately, no copies of the photographs mentioned in the above quote are known to have survived, but there is one photograph of Niven with two men in a cave in the Czar Mine (Figure 1.2.2). Presumably, these men assisted him in the recovery of the many hundreds of the specimens during the two weeks he spent in this cave.

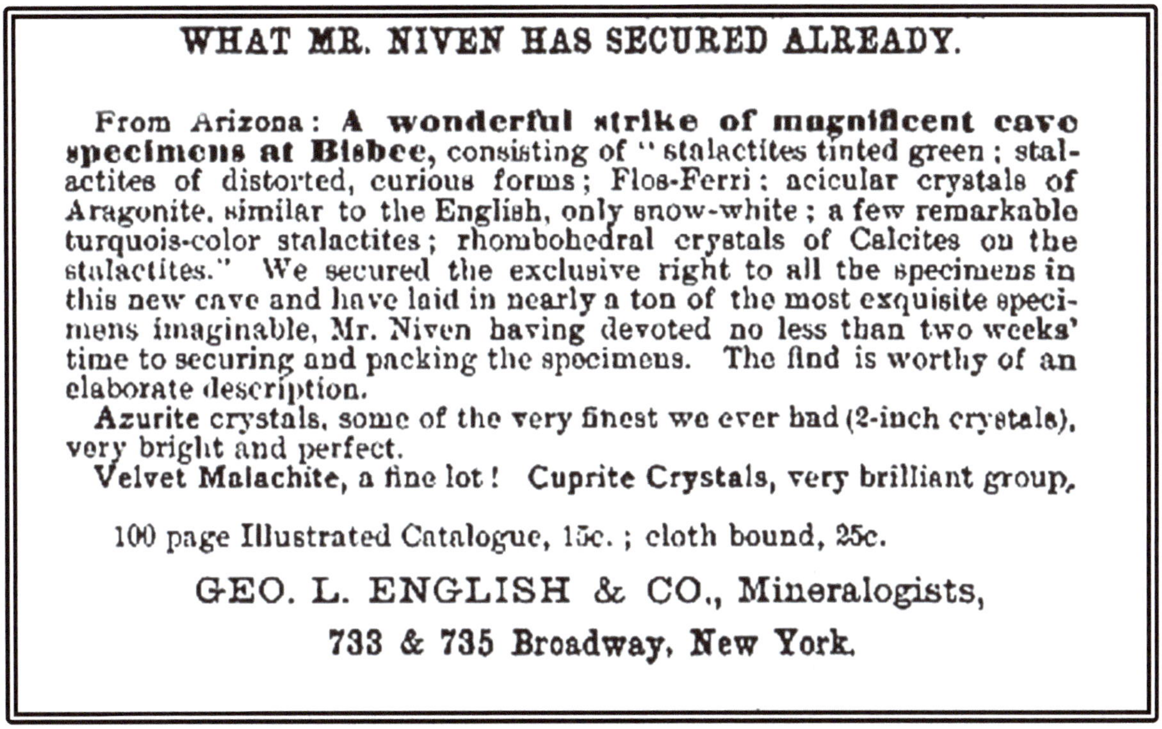

WHAT MR. NIVEN HAS SECURED ALREADY.

From Arizona: **A wonderful strike of magnificent cave specimens at Bisbee,** consisting of "stalactites tinted green; stalactites of distorted, curious forms; Flos-Ferri: acicular crystals of Aragonite, similar to the English, only snow-white; a few remarkable turquois-color stalactites; rhombohedral crystals of Calcites on the stalactites." We secured the exclusive right to all the specimens in this new cave and have laid in nearly a ton of the most exquisite specimens imaginable, Mr. Niven having devoted no less than two weeks' time to securing and packing the specimens. The find is worthy of an elaborate description.

Azurite crystals, some of the very finest we ever had (2-inch crystals), very bright and perfect.

Velvet Malachite, a fine lot! Cuprite Crystals, very brilliant group.

100 page Illustrated Catalogue, 15c.; cloth bound, 25c.

GEO. L. ENGLISH & CO., Mineralogists,

733 & 735 Broadway, New York.

Figure 1.2.3: Excerpt from an 1891 George L. English ad for Bisbee cave specimens (after *The American Journal of Science*, 1891).

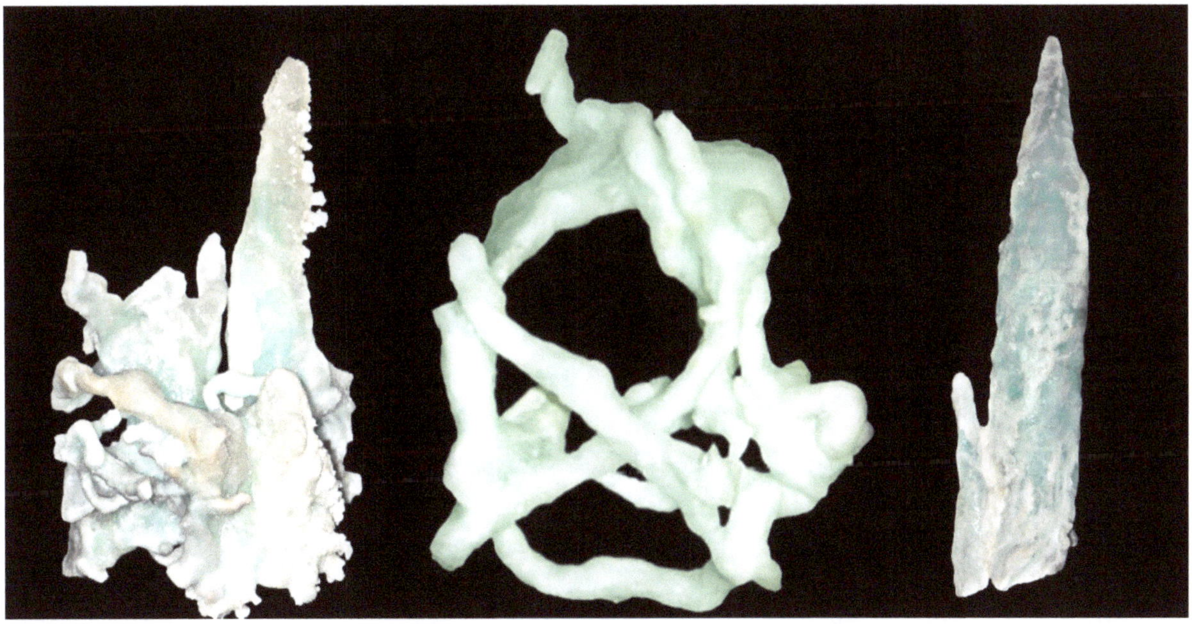

Figure 1.2.4: Copper tinted cave formations collected by Niven in the Czar Mine in 1890 and sold by George English.
10.5 inches (left) 11.2 inches (center) 9.1 inches (right)

In all, Niven shipped some 29 boxes to George English, containing 1,870 pounds of specimens from the cavern. These were sold through George English's mineral business and were much in demand, as nothing like these specimens had been discovered before.

The 1880s through the early 1900s were a time of worldwide industrial fairs and exhibitions. Most

Figure 1.2.5: Czar Mine with smelter (left), and the Copper Queen Mine (center right), 1888.
This is very much like what Niven saw when he arrived in Bisbee in late 1890, a rough, frontier mining camp.

states and territories, even other countries, would send displays of their natural resources and examples of their industrial might to these important events. Mineral specimens from Bisbee were exhibited at most of these fairs during this time, both in the U.S. and in Europe. This began with the Denver Exhibition in 1882 where both the Copper Prince Mining Company and the Copper Queen Mining Company were well represented with the latter displaying a specimen that weighed 1,150 pound and assayed 27.8 percent copper (Chicago Daily, 1882), (Reno Evening Gazette, 1882). Then, in 1884 the New Orleans Exposition received a remarkable silver specimen from the Copper Queen Mining Company (Engineering & Mining Journal, 1884). This presentation of Bisbee ores and minerals continued at least until the Panama-Pacific International Exposition at San Francisco in 1915. However, the most impressive collection of Bisbee minerals ever assembled was displayed at the World's Columbian Exposition of 1893, held in Chicago. The "World's Fair," as it was called, was a magnificent event with more than seven million people attending. All of the states and territories in the Union had sent examples of their best to demonstrate their natural wealth and economic prowess.

Arizona provided a huge mineral display, consisting of several hundred specimens that were placed in the Mining and Metallurgy Building. Appropriately, the Bisbee specimens were at the center of this stunning display (Figure 1.2.6). Arizona desperately needed to impress Americans with its mineral wealth, as it was struggling to achieve statehood, alone and independent of New Mexico. The importance of the role Bisbee minerals would play in Arizona's ultimately achieving statehood cannot be over stated. The Copper Queen Consolidated Mining Company, long a proponent of statehood free of New Mexico, understood this very well and began preparing for the event early.

Several years before the 1893 Chicago World's Fair, the Copper Queen went to extraordinary lengths make sure that the Bisbee portion of the Arizona exhibit was outstanding. Mineral specimens of exceptional size and quality were collected specifically for the Fair. Because of the importance of this fair, these efforts were published in national newspapers. The identical article about collecting of an enormous block of azurite (blue copper carbonate) and malachite from the 200 level of the Czar Mine appeared in both the *Chicago Daily Tribune* in its March 13, 1891 edition and the *Los Angeles Times* on March 30, 1891 as the following:

> *"The Copper Queen Company will exhibit at the World's Fair a mammoth specimen of ore from their mines. The work of chiseling it out has been going on for some time and great care is being taken in its extraction. It is estimated that when ready for shipment it will weigh five tons and be in the shape of a brick. The specimen is from the big stope from which such beautiful specimens have been taken and will, without doubt, be the most attractive specimen on exhibition"*

A number of specimens for the World's Fair were also collected from oxidation caves. While collecting samples of ores with the renowned economic geologist M. N. Bateman in 1914, a miner recounted that he had recovered specimens for the World's Fair from several caves. He mentioned that, in one case, he entered the cave easily by crawling through a small hole, but when he attempted to crawl out, he met an unexpected problem. Sharp calcite crystals were oriented such that the crystals pointed into the cave. As he tried to get out, the sharp crystals tore his clothes and cut him, making the exit both difficult and painful (Bateman & Murdoch, 1914). The authors have had similar experiences and can attest to ease with which such crystals can deliver painful cuts.

Bancroft, in his *The Book of the Fair (1893)* described the Arizona mineral display at the World's Fair as:

> *Arizona's exhibits, adjoining the Colorado section, are displayed to excellent advantage on a raised platform, in the center of which is a monument of copper*

ore, in rich colors of blue and green, one of the specimens which it is composed weighing nearly 7,000 pounds, and the smallest exceeding 800 pounds. Around it are cases of cuprite, azurite, malachite, and other minerals of brilliant hue, some of the samples from the Holbrook mine [sic], where there is a cave of stalactites, being covered by incrustations of silver" (Bancroft, 1893a).

An anonymous author wrote a bit more about the cave minerals on display in the April 24, 1893 edition of the *Chicago Daily Tribune,* noting:

In the Holbrook copper mine [sic] is a cave 1,200 feet long in which has been found remarkable formations of stalactite and stalagmite, several specimens of which give variety to the collection of copper minerals in the various show cases. "

Figure 1.2.6: The Arizona mineral exhibit in Mining and Metallurgy Building at the 1893 World's Columbian Exhibition in Chicago. Specimens collected from the Bisbee caves were among the numerous Bisbee minerals on display. The large block of azurite and malachite from the Czar Mine is in the tall, center case, (*The Engineering and Mining Journal, 1893*).

Bancroft, (1893b) also noted that Arizona had an additional, small display of minerals in the Arizona – New Mexico - Oklahoma, Territorial Building. The Territorial Building display was a small duplication of the exhibit in the Mining and Metallurgy Building. This second display contained a number of cave specimens from Bisbee, which had been collected by Niven in 1890.

Following the World's Columbian Exposition, many of the Bisbee specimens on display in the Mining and Metallurgy Building were donated by the Copper Queen to museums throughout the US and Canada. Phelps Dodge & Company, the owner of the Copper Queen Consolidated

Mining Company, retained a good number of fine specimens for a display in their offices in New York.

The huge block of azurite and malachite was presented to the American Museum of Natural History in New York where, today, it is the centerpiece in the mineral hall, as shown in Figure 1.2.7.

Figure 1.2.7; Right, the five foot tall, four ton block of azurite and malachite removed from the 200 level of the Czar Mine in 1891 for exhibition at the 1893 World's Columbian Exposition, as it now appears in the American Museum of Natural History in New York.

During the 1880s and 1890s, it was common to find superb specimens of several other mineral species in and near the oxidation caves. Many of these specimens occurred in stalactitic forms. Malachite was by far the most common of these while goethite was the second most abundant.

Stalactites of azurite were rare and typically occurred within azurite and or malachite/azurite masses in the orebodies below the caves and not in the oxidation caves, per say. Just as the calcite and aragonite from the oxidation caves were in great demand, so were these malachite stalactites.

Figure 1.2.8: Above, goethite stalactites with minor malachite and halloysite, Copper Queen Mine, view – 6.5 inches.

Figure 1.2.9: Calcite with minor aragonite, Czar Mine, Bisbee Mining and Historical Museum specimen, view 4 inches.

Figure 1.2.10: A group of malachite stalactites, 15 inches high, covered with malachite pseudomorphs after azurite and associated botryoidal halloysite, recovered from a cave in the Holbrook Mine.

Cave formations from Bisbee, particularly those tinted by copper remained in demand for many years. Thus, they were collected and sold by a number of mineral dealers with George English advertising in 1896:

*"**Copper stained stalactites** from Bisbee are the most beautiful ever found. We have a new and large lot of splendid specimens"* (The Mineral Collector, 1896a).

Some of the larger stalactites were sliced and the polished slabs sold by other dealers. These slabs are quite scarce today and remain highly desirable, as do all of these early specimens.

Figure 1.2.11: Right, polished, copper tinted, calcite stalactite, labeled Bisbee, Arizona, view - 6.5 inches, USNMNH specimen.

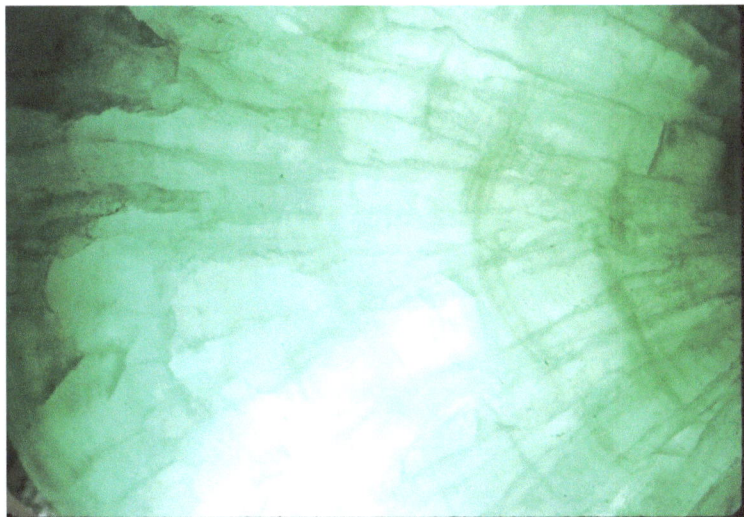

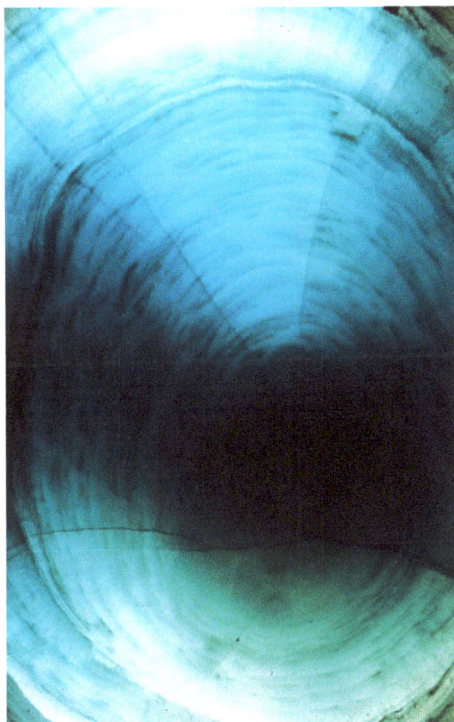

Figure 1.2.12: Polished, copper tinted aragonite stalactite, Czar Mine, view - 10 inches, USNMNH specimen.

Figure 1.2.13: Top ad – The Mineral Collector – 1896b.

Figure 1.2.14: Bottom ad – The Mineral Collector – 1898.

It is worth noting again that by allowing, if not encouraging commercial collecting, Dr. Douglas demonstrated great foresight in fostering the preservation Bisbee's mineral specimens. Douglas and others after him at the Copper Queen often went to extraordinary lengths to save the wonderful specimens found in the mines and allowed numerous visitors to see the beautiful caves before they were to be mined. As a result, today we have beautiful and scientifically valuable mineral collections from many of the oxidation caves, as well as written records and a small number of photographs of these wonders.

In a few instances, the Bisbee caves were particularly rich with copper minerals, particularly malachite and, to a much lesser degree, azurite. Perhaps the most important and prolific azurite occurrence ever found, anywhere in the world, was in an oxidation cave at Bisbee, yet it has gone unrecognized as such. It was also one of the most unusual caves found anywhere. This small cave in the Southwest Orebody of the Czar Mine, hit in early 1895 was described as:

> *"A room, not too big, perhaps 50 feet in curved length and 20 feet high and 15 or so feet wide. The walls were all manner of irregular lumps of black azurite dotted with malachite. From the back* [ceiling] *hung limonite stalactites with azurite crystals here and there on them. The floor was mostly a thin crust of blue (azurite) on malachite"* (Graeme, 1981).

No doubt, hundreds of fine and often very large azurite specimens were collected from this cave in the Southwest Orebody of the Czar Mine. Numerous azurite specimens exceeding several feet across from this cave continue to grace collections worldwide. Fine, smaller pieces are relatively abundant, yet highly treasured by their owners everywhere.

This cave occurred in massive, impure goethite on the lower edge of the very large Southwest Orebody. The opening most probably formed along a marginal crack related to oxidation subsidence of the encasing limestone. Additional goethite was deposited within the cave as the ore was oxidizing and it occurred as typically small stalactites and larger botryoidal forms. In the lower portions of the cave, were collapse material accumulated, were smallish pieces of impure goethite, which had fallen from the cave walls and ceiling during oxidation and cave formation. All served as a base for azurite.

Azurite was deposited on the cave walls and floor, as well as, for some distance into the porous, brecciated goethite wall rock. From a study of numerous specimens recovered from the cave, it is apparent that it was filled with copper rich solutions, which deposited azurite, at least twice during the cave development process with a period in between when it was air filled. Both episodes of azurite deposition occurred in a subaqueous (below solution level) environment only, as no specimens reflecting azurite formed in subaerial (open air) depositional environment are known from this particular locality.

Azurite crystals of up to three inches were the first to be deposited in what became the upper part of this cave. These were subsequently altered to malachite (pseudomorphs of malachite after azurite) in an oxygen rich, subaerial environment. At the time when the cave was air filled, malachite was deposited in the lower parts of the cave as irregular masses to several inches thick.

Figure 1.2.15: Azurite; second-generation overgrowth on malachite pseudomorphs after azurite and malachite on goethite, Czar Mine, view - 3.75 inches.

Figure 1.2.16: Malachite from the cave floor showing the thin layer of goethite staining followed by a thin azurite crust on top and tiny cuprite crystals, Czar Mine, view – 3.4 inches.

Figure 1.2.17: Azurite overgrowth on goethite stalactites, specimen – 8 inches, Czar Mine.

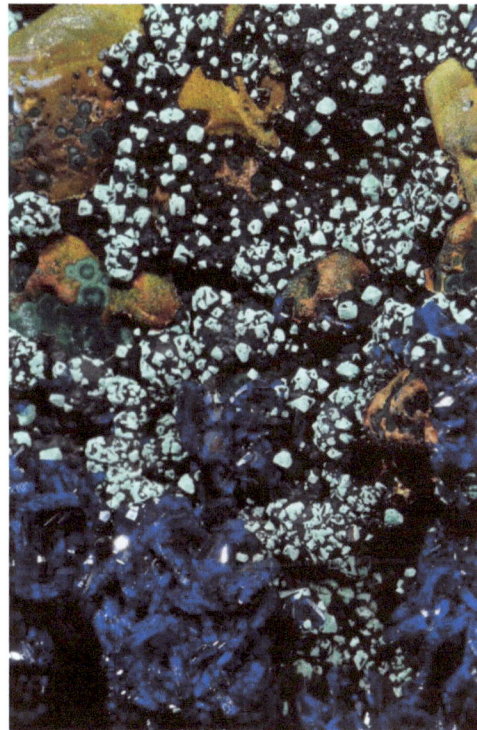

Figure 1.2.18: Above, malachite coating cuprite on azurite and on goethite coated malachite, view – 2.5 inches, Czar Mine.

This malachite was exceptionally lustrous and chatoyant as illustrated in Figure 1.2.16. At some point following the malachite deposition, the lower portions of this cave appear to have been submersed with low pH, iron rich solutions. A thin veneer of goethite or at least surface alteration or staining from the iron is present on all of the malachite as is evidence of minor corrosion of the malachite by these mildly acidic solutions.

The second period of azurite deposition appears to have been responsible for the majority of the azurite found in this cave. Two forms of azurite deposition are apparent; one form is second generation over growths on the existing malachite pseudomorphs after azurite (Figure 1.2.15). The other is a composite form and is the dominate type. In this case, azurite occurred as open spherical clusters of small, tabular crystals (see Figure 1.2.19).

Interestingly, not one of these distinctive crystal groupings occurred as overgrowths on the earlier malachite pseudomorphs after azurite. Rather here, a thin veneer of azurite was deposited in an oriented fashion on these large pseudomorph crystals. Because this layer of azurite was quite thin, a much lighter blue color than would normally be seen in a large azurite crystal resulted.

Lastly, small amounts of tiny cuprite octahedral crystals were deposited on some the azurite, malachite and goethite. These tiny, scattered crystals are invariably altered partially or completely to malachite as shown in Figure 1.2.18.

Figure 1.2.19: Azurite with minor malachite as coatings on cuprite on goethite, Czar Mine, specimen - 10.6 inches.

Fine examples of cave speleothems were found in another cave associated with this same orebody at much the same time and recovered for sale. All of these were noted by Bates in the February 1896 issue of *The Mineral Collector,* as shown in Figure 1.2.20 below.

⟶ THE ⟵

MINERAL COLLECTOR.

[Entered as Second-Class Matter at the New York, N. Y., Post Office, March 19, 1894.]

| VOL. II. | FEBRUARY, 1896. | No. 12. |

News and Comments.

By Albert C. Bates.

BISBEE, ARIZONA.—The Copper Queen Mine, famous as the place of occurrence of the largest masses of velvety malachite ever found, some of the finest of which are in the American Museum of Natural History, in 1895 yielded a large quantity of extraordinary azurites, malachites, cuprites and calcite-aragonite stalactites.

The azurites occur on a musty brown matrix in masses of sharp crystals, or in isolated well defined crystals (some of which are doubly terminated), and on a hardened malachite in drusy crystals. Some quite large crystals are coated with, or have altered to, malachite.

Indeed, it is common experience to discover azurites from this locality partly changed to malachite after being in a cabinet for years.

Some specimens are sprinkled over with minute green crystals of Cuprite; others show the most delicate tufts of malachite, and some show all of these minerals in harmonious combination.

The stalactites were found in fantastic fringes, in pipe-stems, in bunches of coraloidal forms and gnarled masses, from pigmies to giants in size. Some are snow white, some a delicate green, others white sprinkled with green dots, and some are of a silken sheeny lustre. It is reported that but one pocket was found. Collectors should take advantage of offerings at once.

Figure 1.2.20: Note on the occurrence of the described minerals in the *Mineral Collector,* (Bates, 1896).

Today, these many fine specimens of azurite, malachite and calcite from the oxidation caves associated with the Southwest Orebody still occupy places of honor in collections throughout the world.

Figure 1.2.21: Right, azurite and minor malachite on goethite from the occurrence and of the type described by Bates (1896), Czar Mine, specimen - 2.5 inches.

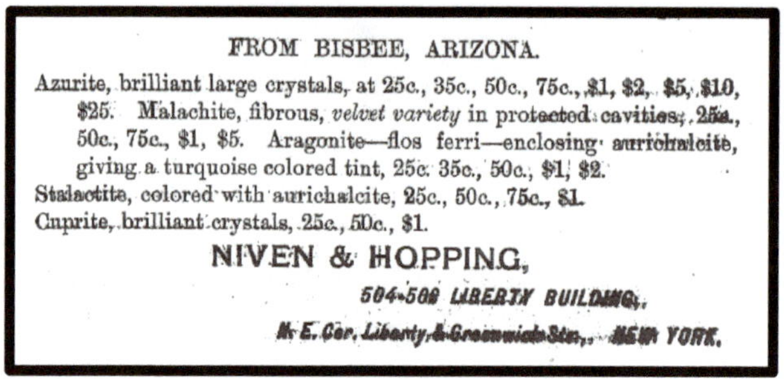

Figure 1.2.22: Right, advertisement for Bisbee minerals after *The Mineral Collector*, (1896b).

FROM BISBEE, ARIZONA.

Azurite, brilliant large crystals, at 25c., 35c., 50c., 75c., $1, $2, $5, $10, $25. Malachite, fibrous, *velvet variety* in protected cavities, 25c., 50c., 75c., $1, $5. Aragonite—flos ferri—enclosing aurichalcite, giving a turquoise colored tint, 25c. 35c., 50c., $1, $2. Stalactite, colored with aurichalcite, 25c., 50c., 75c., $1. Cuprite, brilliant crystals, 25c., 50c., $1.

NIVEN & HOPPING,

504-508 LIBERTY BUILDING,

N. E. Cor. Liberty & Greenwich Sts., NEW YORK.

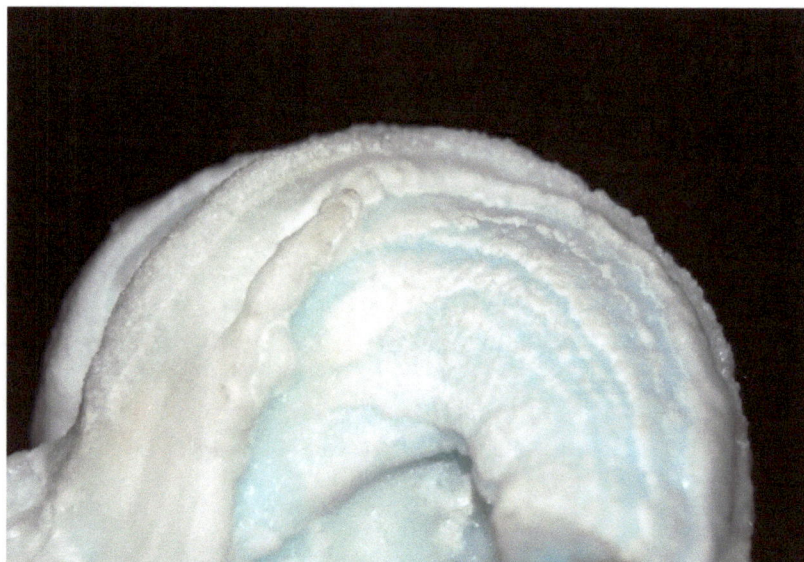

Figure 1.2.23: Left, aragonite tinted "a turquoise color" of the type noted in the above ad, Czar Mine, view - 3.6 inches.

31

During October 1896, Arthur Curtiss James stopped by Bisbee on his way to New York as he returned from a sailing trip to Japan. James, a descendent of the Phelps family, was one of the largest shareholders of the closely held Phelps Dodge and Co., the owners of the Copper Queen Consolidated Mining Company. His journey was chronicled by lady member of the traveling party. Upon arrival at Bisbee, they were met by Ben Williams, superintendent of the Copper Queen Mine and in charge of all operations at Bisbee. No doubt Mr. Williams was most eager to extend every courtesy to someone of such importance in the company. While in Bisbee, they made a trip underground and visited the cave in the Southwest Orebody, referenced above, even though it had been extensively mined by this time. Here is how she recorded their first hours in Bisbee and the trip into the mine:

"Bisbee looked very natural, although we find the works far more extensive than five years ago. Mr. Ben Williams met us and after coming to the car for a few minutes, we went to his house, where we found Mrs. Williams and the rest of the family. Mr. Williams lunched with us, and we arranged for an afternoon trip into the mine. The men clothed themselves in blue jeans at the store and we donned wash dresses. We visited the two hundred and four hundred foot levels, and walked about two miles underground, climbed into one stope fourteen feet below the two hundred foot level, and saw one chamber of the beautiful blue and white cave discovered some time ago, the rest of which is now unsafe. Much of the blue and green ore has been taken out, and that now being worked, although rich with copper, is less beautiful. We were all covered with mud and soil of various colors when we came to the surface again, and the two bathrooms in the Williams' house were in great demand" (Coronet Memories, 1899).

Figure 1.2.24: Dirt smudged, young visitors in a small cave in the Czar Mine, C – 1896.

Undoubtedly, the most widely published description of a visit to a cave at Bisbee occurred when a number of national and international newspapers and magazines reported on an 1897 meeting of the Masonic Grand Lodge of Arizona, which was held in a spectacular cave on the 200 level of the Czar Mine. A pre-meeting article in the November 3, 1897 issue of *The Arizona Republican* reported:

"The cave at Bisbee where the Masonic Lodge will meet"

"Mr. Lewis Williams of Bisbee, better known there as Don Luis was in Phoenix yesterday morning. Mr. Williams says regarding the approaching meeting of the Masonic Grand Lodge there, that letters have been received from Masons in all parts of the United States regarding the meeting room, a cave in the Copper Queen Mine. Many members of the order to whom neither distance nor expense is significant will be present, attracted by the most unusual meeting place. It is the most beautiful known cave which lies between the fourth and fifth levels of the great mine and is arrived to from the latter level. It is of great extent and a height varying from two or three feet to thirty feet. It is wholly covered with a lime deposit. The deposit on the floor bears an exact resemblance to running waters and the course of the currents. Stalactites vary in size from the tiniest icicle to the thickness and stature of a man, dependent from the ceiling and are met by stalagmites. Sometimes the lime deposit takes the form of stretches of the finest lace and is so fragile that it shivers at a touch. A beautiful effect is produced in the cavern by placing candles behind this lace work. On the occasion of the grand lodge meeting the cavern will be lighted by electricity, the globes of the incandescent lamps being of many colors. What became of the water by which, in the course of ages, this great cave was leached out was at first a mystery. A year or two ago a hole was broken through a wall of the cavern, disclosing another cave, which has never been explored and perhaps never will be. It is a horrible fissure, not quite perpendicular. Pieces of stalactites or rocks thrown into it, bound from side to side until the sound dies away in the immense depth. Sometimes when an object is thrown into the chasm follows its course without striking the sides, no sound is heard. It is as if the object had been thrown into limitless space. It was no doubt through this chasm that the waters fled, where, no one knows, after carving out this grand Masonic lodge."

Lewis Williams was the brother of Ben Williams and superintendent of the Copper Queen Smelter at the time and, as members of the Bisbee Masonic Lodge, the Williams brothers were the host for this important meeting. The reported location for this cave was *"between the fourth and fifth levels of the great mine"*. The fourth level of the incline, Copper Queen Mine was equal with the 200 level of the Czar Mine, through which the Masons would enter. The deep fissure noted in the article was undoubtedly one of the oxidation related subsidence cracks, so frequently associated with these caves. These cracks were often several hundred feet in depth. Such a depth is sufficient to completely muffle the sound of a small object thrown into such an opening.

The work to prepare the cave for this prestigious meeting must have been substantial. These openings are seldom easy to traverse, much less accommodate a solemn and stoic meeting of several hundred people. So unusual and so special was the meeting that a brief note about the cave and the necessary preparations can be found in the Arizona Territory section of the *Annual Reports of the Department of the Interior* (1898), wherein it states:

"The cave lies in a northeasterly from where the mine is entered, and is said to be about 900 feet under the surface of the hill in which it is situated. It is probably nothing more than a great bubble in the limestone formation. So far as can be judged by observation, it is probably 250 by 300 feet in extent and 65 feet deep in the center, although much of the depth has evidently been lost by the falling of stalactites, some of them weighing tons, and which because of their great weight had broken from the roof; others looked as though they needed but little encouragement to do the same thing. The west end had been partially filled and a platform capable of seating 300 people erected thereon, and the whole brilliantly illuminated by scores of incandescent lights. In the letter "G" suspended in the east, no fewer than thirty-two electric bulbs had been placed. On the platform and extending into the cave the electric wires had been shaped into a square and compass. The extended points of the compass were 100 feet apart and the shaft of the square 120 to the angle. In the formation of the mammoth emblem of the masonry 56 electric lights were used, and numerous other lamps were placed elsewhere about the cave. An idea of the magnitude of this work may be gathered in the fact that $3^1/_2$ miles of wire were used in it" (Department of the Interior, 1898).

Figure 1.2.25: The cave and facilities in the Czar Mine just before the meeting of the Arizona Masonic Grand Lodge in November 1897, (Craft, 1899).

The meeting of the Grand Lodge must have been an impressive event for the small, remote community of Bisbee. The procession of November 10th to the very industrial appearing and less than attractive surface facility (Figure 1.2.26) which lay over the spectacular meeting place, some 200 feet below, was described as:

"Clothed in white gloves and aprons the two hundred men in line made an imposing appearance as they marched from the Lodge to the hoisting works from which they were to descend into the mine and cave. In the line of the march the Grand Lodge were in the rear, but on reaching the works the column halted, opened ranks and the Grand Lodge passed through, and were, of course, the first to enter the mine. They were scientifically stood on the cage a half a dozen at a time, when down they went about two hundred feet in a second to the level by which the cave was reached. Then began the long march in single file through the winding, angling tunnel, which was illuminated the entire distance by candles set about ten feet apart, and each turn and crosscut carefully guarded by Masons who are employed in the mine, till at length the individual passed through a temporary door into a scene of inexpressible splendor and beauty – the illuminated cave" (Grand Lodge of Arizona, 1897).

Figure 1.2.26: The Czar Mine hoisting works (center-left with white window frames) and the massive Copper Queen smelting works, as seen by the Masons during their visit in 1897.

Though admittedly *"a scene of inexpressible beauty and splendor,"* an effort was made to describe the cave in the most flowery language as presented below:

> *"The figure, the strength, the splendor, the symmetry, the polished alabaster, the glittering wealth of gold and jewels of King Solomon's temple* [sic] *were here dazed and dimmed into insignificance by the brilliance, the abounding wealth of exquisite forms of beauty, the majestic splendor of huge abutments literally covered by the most delicate tracery, the flash of a million jewels; the bold outline forms of grandeur and strength clothed in glittering delicate chiseling, such as no human hand has ever wrought, the stupendous dome of the mountains above borne upon those noble and beautiful pillars, and all fresh from the hand of the Supreme Grand Architect of the Universe, filled one with the overwhelming emotion of surprise, of delight, of adoration.*
>
> *Stalactites from the size of a drop of water to that of the great tusk of some antediluvian monster hung suspended from the roof, and beneath the white glare of*

Figure 1.2.27 Meeting of the Arizona Grand Lodge of the Masons in the Copper Queen Cave on the 200 level of the Czar Mine, November 10, 1897.

the electric light they danced and shimmered like icicles in the sun. Singly and in cluster, some in blue and some in white, of all lengths and shapes, these stalactites cover the roof, while among them in the labyrinthine irregularity glitters the crystal ooze.

On the north side, almost opposite to where the cave was first broken into the line, God has created the fairest creatures of his handiwork, and what King Solomon wrought in years was here fashioned in a single night; but a night that knew no day, nor the sound of ax, hammer or toll of iron till the operative workmen, agreeably to the designs drawn upon the trestle board, reveled its hidden treasures. Here are the steps, the pot of incense, the bee-hive, the hour-glass and water fall. The steps, whiter than Parian marble, from great coils of alabaster rope, are from two to ten feet in diameter and six feet high. East of these is a waterfall, stayed midway in its decent by some hypnotic hand; white and awe inspiring in its grandeur is this seemingly stilled torrent of foaming waters that ere it slept had dashed its spray on hammocks of ice in a thousand fantastic shapes sparkled in the light. There are curtains and veils behind which no man may enter, woven white and transparent in the Cimmerian darkness. The whole aspect of the cave is one of enticing and bewildering loveliness, and he who can gaze without awe upon the grand scene has no beauty in his soul.

The cave, which is probably one hundred and fifty by two hundred and fifty feet in dimensions, and an extreme height of about one hundred feet from the lowest to the highest point, was lighted by electricity. Eighteen thousand feet of wire used, some of them one hundred candle power. In the east hung a brilliant letter "G," three feet in height, and studded by thirty-five electric lights of sixteen candle power, a thing of great beauty and joy in itself. In the vast space overhead the lights formed a great square and compass, the distance between the points of the extended compass being one hundred and thirty feet. A sufficient portion of the cave for the use of the Masons had been floored over and seated, and there beneath that great dome, seventy feet overhead, and surrounded by such glitter, such beauty, such strength, such overwhelming sense of location, such nearness, and such a dependency upon the immediate handiwork of the Grand Master Mason of all, the Grand Lodge of Arizona was called on by the M. W. Grand Master" (Grand Lodge of Arizona, 1897).

GRAND LODGE OF ARIZONA A. F. AND A. M.

Held in the wonderful Copper Queen Cave at Bisbee, Arizona, Dec.9th to 13th, 1897.

As Taken by Flash-Light.

THIS picture shows the Grand Masonic Lodge of Arizona as it was held in the wonderful cave in the mine of the Copper Queen Consolidated Mining Company. The first and only time in modern Masonic history, where this illustrious order held its meeting in one of Natures most beautiful lodge rooms. It is reached by decending a perpendicular shaft, two hundred feet thence through a winding tunnel over eighteen hundred feet long. The cave is nearly two hundred and fifty feet in length, one hundred feet in width and nearly the same in height, is almost entirely covered with pure lime crystals, some of which hang in great clusters resembling icicles, others on the sides beautiful frozen waterfalls, domes and towers, in color they are principally white, but in places have the beautiful green copper stain. As seen in the picture, the cave was beautifully illuminated by elect.ic light, (the letter "G." being entirely composed of incondescent globes) it presented a most wonderful and dazzling appearance,a scene never to be forgotten by those who witnessed it. Below appear the names and residences of the brothers who were present.

Figure 1.2.28: The above is a portion of the information that was printed on the back of the cardboard on which the above group photograph, Figure1.2.27, was mounted (Miller, 1897). The December dates on this card conflict, by one month exactly, with the November dates officially recorded in the Proceedings of the Grand Lodge of Arizona (1897).

A number of photographs of this unforgettable event were taken by Mason and photographer Andrew Miller from Globe, Arizona who attended this very special event. Miller offered these photographs for sale as souvenirs. The Christmas Day, 1897, edition of the *Arizona Daily Star* reported that a Tucson attendee of the meeting had several of the photographs in hand.

> *"Kirk Hart has received a number of photographs of the underground Masonic Lodge of Bisbee, taken by flash light during the recent session of the Grand Lodge at that place. There are three different pictures. One of the cave room with the furniture, etc. ;*(see Figure 1.2.25) *one with the members of the Masonic Grand Lodge of Masons of Arizona, its officers and visiting members of the Blue* Lodge (see Figure 1.2.27). *This is very good as nearly all in the picture can be recognized. The most distinct is that of the Grand Commandery* [sic.] *where the group of officers and members is not so numerous and could be focused to better advantage. Every Knight is easily identified. Many of the striking features of the cave are brought out in these pictures. The work of the artist is highly creditable, and a most beautiful souvenir of the most interesting congregation of Masons ever held in Arizona."*

Keeping with the long tradition of permitting many to see these natural wonders, the managers of the Copper Queen allowed more than just their Masonic brothers to see this cave. According to *The Arizona Republican* edition of November 14, 1897, hundreds of visitors were allowed into the cave to view its splendor. Also, Craft, (1899) noted:

> *"The cave has since been seen by people from many places in the world..."*

This great cave collapsed just months after the memorable event. Notice of this cave-in was given, by *The Arizona Republican* on February 13, 1898 with the following brief statement, which was originally published by the *Tucson Star*:

"It is learned from a gentlemen who came in from Bisbee that the beautiful stalactite cave at that place had fallen in last week. No one was injured however."

Figure 1.2.29: Queen Hill above the Czar Mine with a few of the numerous subsidence cracks reopened by mining activity noted by arrows - 1909. Local lore incorrectly has it that these cracks were the result of a significant earthquake hit the area in 1887.

That no one was injured was an extremely important note. Cave-ins were an ever present danger in all mines and a concern for everyone underground, most of all in the heavy ground associated with the ores at Bisbee.

The collapse of such a large opening was an extremely hazardous event, one that would potentially affect all nearby areas by destabilizing the ground for hundreds of feet. And too, a sudden, catastrophic collapse of this big hole could have caused what miners call an "air blast," forcing large volumes of air through the mine opens at high pressure, with the potential to injure miners well outside the cave. This does not appear to have happened, but it was one of the hazards associated with large, open spaces.

No doubt, the highly fractured rock, which had naturally developed above the cave during its formation, became unstable because of the mining of the ores below and nearby contributed to the collapse. More than one such collapse of open spaces occurred over the years.

Often, the cracks related to a collapse extended to the surface. Subsidence due to ore removal was responsible for most cracks. Many of the subsidence/collapse cracks, which can be seen in Figure 1.2.29, are still obvious today on Queen Hill above the Queen Mine Tours.

Interestingly, there has long been the erroneous belief among many Bisbee residents that these cracks were the result of the earthquake which shook Bisbee and much of southern Arizona in May 1887. DuBois and Sbar, (1981) estimated the magnitude of the quake to have been 7.25.

James Douglas was coincidently in Bisbee at the time, and did not report the development of any cracks in Queen Hill. The event was detailed in a paper he coauthored with his long time friend and colleague, Thomas Hunt (Hunt & Douglas, 1887), wherein much information concerning the quake and its affects was presented. It is inconceivable that he would not have reported cracks developing in Queen Hill, given his penchant for detail.

Further, a note in *The Engineering and Mining Journal* (An Occasional Correspondent, 1887) described the quake noting that rocks were dislodged from the steep hillsides and of very minor structural damage to a building, but no mention was made of cracks developing in the surface because of the earthquake. This absence of a reference to cracks forming is telling in that the article went into great detail to discuss the effects of this seismic event on the surrounding areas. Thus, it is safe to assume that none of the cracks were a result of this earthquake, something borne out by post-quake photos such as Figure 1.2.5.

Chapter Three: The 1900s

The first concerted effort to present a geologic overview of the Bisbee ore deposits was made by Dr. James Douglas in 1899 when he delivered a paper to the New York meeting of the American Institute of Mining Engineers. His paper, with illustrations, was printed in 1900 in the very highly regarded *Transactions of the American Institute of Mining Engineers*. In this paper, Dr. Douglas correctly suggested that the ores had been deposited as sulfides replacing limestone, a change from his earlier position on the subject. Douglas also shared his belief that the caves had formed because of sulfide oxidation and the related shrinkage of the residual material.

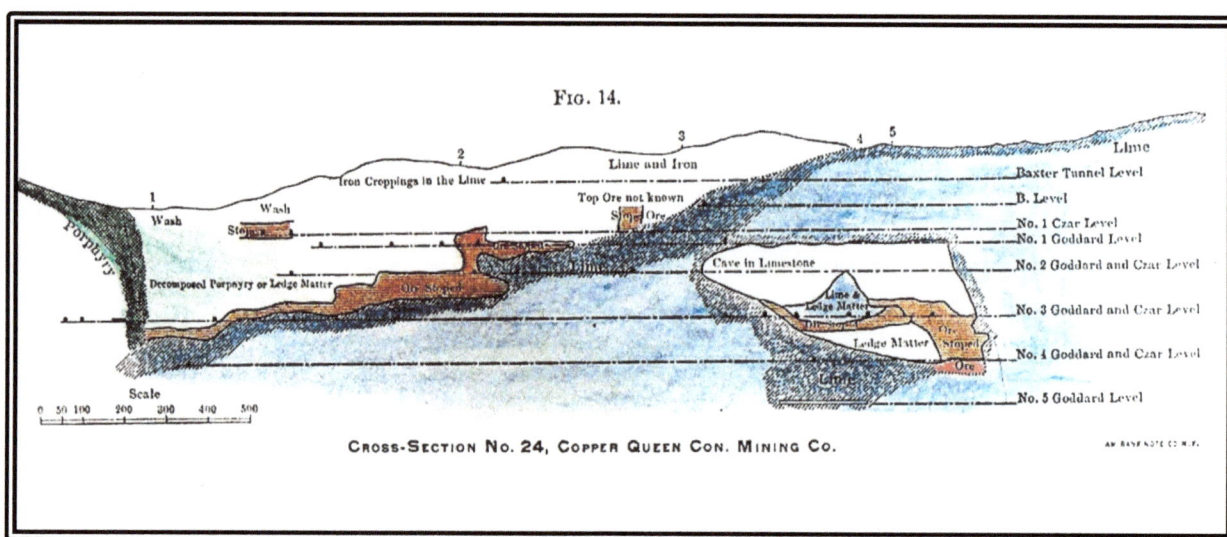

Figure 1.3.1: Cross section showing a very large cave in the Holbrook (Goddard) Mine over ore. (Note scale at the lower left and be advised that the levels in the Czar Mine are 100 feet apart), after (Douglas, 1900).

This statement on the origin of the caves was a revolutionary thinking at the time and one, which even today, is not universally accepted. In his paper, Dr Douglas wrote:

> *"A feature of the Bisbee mine is the large caves, which had some influence on the occurrence of the oxidized ore bodies. The walls, roofs and floors of these caves are always covered with stalactitic accretions, which are often tinted green, blue and red by the copper and iron solutions which are mixed with a solution of lime. What, however, gives these caverns practical interest is that they have invariably covered oxidized ore bodies. Fig. 14* [shown here as Figure 1.3.1] *gives a cross-section through one of the large caves. Three such caves of considerable extent have been encountered, and in every instance this combination has occurred. It may be accidental; but so satisfied are we to the contrary that, when a cave is now met with, drifts are run beneath it to strike the ore body. It is a fair assumption that the cave, if not originally formed by the contraction of an ore-body, was increased by the shrinkage of the latter during its oxidation, and that, therefore, a genetic relation really exist between the cave and underlying ore"* (Douglas, 1900).

Douglas notes that just three large caves had been hit operations in the several great mines of the Copper Queen Consolidated Mining Company. The cave illustrated in Figure 1.3.1 was nearly 500 feet long and more than 150 feet in height, a substantial opening by any standard. Caves of this size were few indeed. However, what Douglas did not mentioned was that a great many relatively small caves had been found during the previous 20 years as well. These too had ore underneath or within, which had been mined.

Figure 1.3.2: Fissure on the 200 level, Czar Mine (Douglas, 1900). This was undoubtedly a subsidence caused, marginal crack.

Figure 1.3.3: Cave in the Czar Mine (Douglas, 1900). Note the man and the ladder at the bottom, and to the left of the large, center stalactite.

By 1900, when Douglas published his paper, the Copper Queen Consolidated Mining Company had expanded its operations much to the east of the Czar Mine. This came with the purchase of the nearby Holbrook and Cave properties, which were developed during the early 1890s as the Holbrook Mine. The location of the Holbrook Miner relative to the Czar is shown in Figure 1.3.4. While the ores in the Holbrook were a bit deeper, many were very similar to those found in the Czar and the Copper Queen mines in that they often had oxidation caves associated with the rich oxide ores. Soon, the Holbrook became the most important producing mine at Bisbee, but it was not the only new mine developed by the Copper Queen during this period.

Not far from the Holbrook Mine was the Silver Spray Mine or Spray, as it was commonly called. Figure 1.3.5 shows this mine. It was developed during the mid-1890s and became an important ore producer for the company. While the ores here were even deeper yet, completely oxidized ores and the associated oxidation caves were also found on the upper levels in the Spray. Most of the truly deep ores in the Spray were incompletely oxidized or sulfides and it was these ores that were largely mined from 1909 until the mine was ultimately closed in 1918 (Mills, 1956).

Also in 1900, the well-regarded economic geologist, James Kemp, wrote about the Bisbee cave/ore association when he stated:

"Above the bodies of ore empty caves are usually found, and so frequent is this association that when the prospecting drifts strike a cave the miners immediately sink in the expectation of striking an ore body in depth. Sink-holes on the surface have been successfully used as guides in the same way."

Figure 1.3.4: Holbrook Mine (center left) with Queen Hill behind, and the Czar Mine and smelter in the center. Bisbee is in the background - 1900, at the time of Dr. Douglas' important paper on Bisbee was published.

When Kemp wrote *"Sink-holes on the surface have been successfully been used as guides in the same way,"* he was actually referencing surface depressions. These had formed due to oxidation subsidence, not an actual opening that led into a cave as the term sinkhole typically is used to denote. There is no record of any true sinkholes at Bisbee.

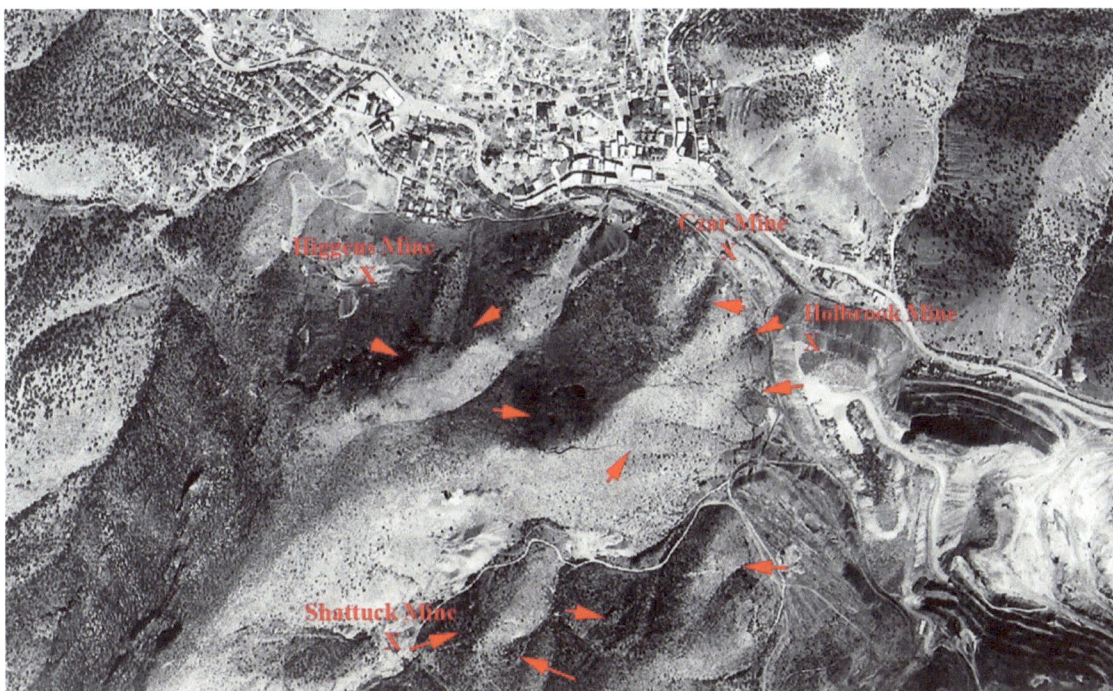

Figure 1.3.6: Aerial view of the Queen Hill area from the Higgins Mine to the west, the Holbrook Mine to the east, the Czar Mine to the north and south to the Shattuck Mine with many of the subsidence depression features indicated, 1958.

However, surface depressions caused by oxidation subsidence were common and their relationship to ore below long recognized. Many are still visible on the hills today (see Figure 1.3.6). These depressions were roughly conical in form and reflect the subsidence of the limestone beds immediately above the oxidized orebodies. Wisser, (1927) discusses these and illustrates their relationship to ore.

Figure 1.3.5: The Spray Mine, C-1910

The impressive stalactites from Bisbee's oxidation caves remained in high demand and popular theme among collectors. An article on stalactites by A. C. Bates in the mineral section of April, 1902 issue of *Popular Science News,* talked about these with the following, accompanied with Figure 1.3.7:

> *"A cavern discovered in the Copper Queen Mine, at Bisbee, Arizona, contained what was considered from a collector's point of view the finest stalactites ever*

found. Delicately fantastic in makeup, some of the stalactites were so thin and broad as to suggest a waving flag, some showed pearly tinting and others varying shades of green, caused by the copper-laden drippings. The removal of these stalactites was made necessary for the further development of the mine, and the best of them are now in museums and private collections all over the world. The New York Museum of Natural History has one of its largest cases filled with these stalactites."

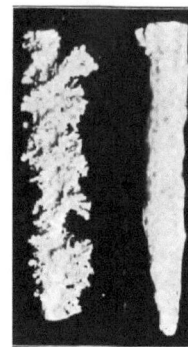

STALACTITES, BISBEE, ARIZONA.

Figure 1.3.7: Stalactites from Bisbee (Bates, 1902).

As a part of documenting the abundant natural resources in America, the United States Geological Survey sent the best economic geologists of the era to study the Nation's most important mineral deposits and to prepare detailed documents describing them for public use. In 1902, Fredrick L. Ransome of the U.S. Geological Survey arrived in Bisbee to study and document the geology of these incredible copper deposits. By the time Ransome arrive in Bisbee, there had been, just over 20 years of continuous mining with no signs of the ore nearing depletion. Indeed, the expansion of mining activity in the Warren Mining District was at an all-time high with a number of new companies now active in the area.

Ransome produced the first, truly comprehensive study of the ore deposits at Bisbee in the context of the regional geology. His final report was published in 1904 as *The Geology and Ore Deposits of the Bisbee Quadrangle, Arizona, U.S. Geological Survey Professional Paper 21.* It was an extraordinary work, one which has stood the test of time. For more than 70 years, Ransome's work was used to guide the search for ore throughout the Warren Mining District. For geologist working at Bisbee, this was, indeed, the "Bible" for their exploration efforts.

While Ransome confirmed Dr. Douglas' opinion that the ores were sulfide replacement deposits in limestone, he wrote very little about Bisbee's caves and nothing of their origin. This was, no doubt in part, because few if any such caves were available for him to visit during the relative short time he was working in the area. When Ransome arrived in Bisbee, most of the ore was being mined at depths below the level of total oxidation. At these levels, the conditions for oxidation cave formation were not favorable.

He did however, write the below generalized description of the caves under his discussion of the occurrence of malachite in the mines:

"The walls of these caverns were covered with velvety moss-green malachite and sparkled with blue crystals of azurite, while from the roofs hung translucent stalactitic draperies of calcite, delicately banded and tinted with the salts of copper" (Ransome, 1904a).

45

Figure 1.3.8: The wall of a cave on the 100 level Holbrook Mine similar to the generalized description given by Ransome (1904a). Malachite and chrysocolla coat the cave wall with later calcite and aragonite partial overgrowth. Much of the calcite is tined green by copper or red-brown by iron while the aragonite is colored a blue-green by copper. Vertical view – 12 feet.

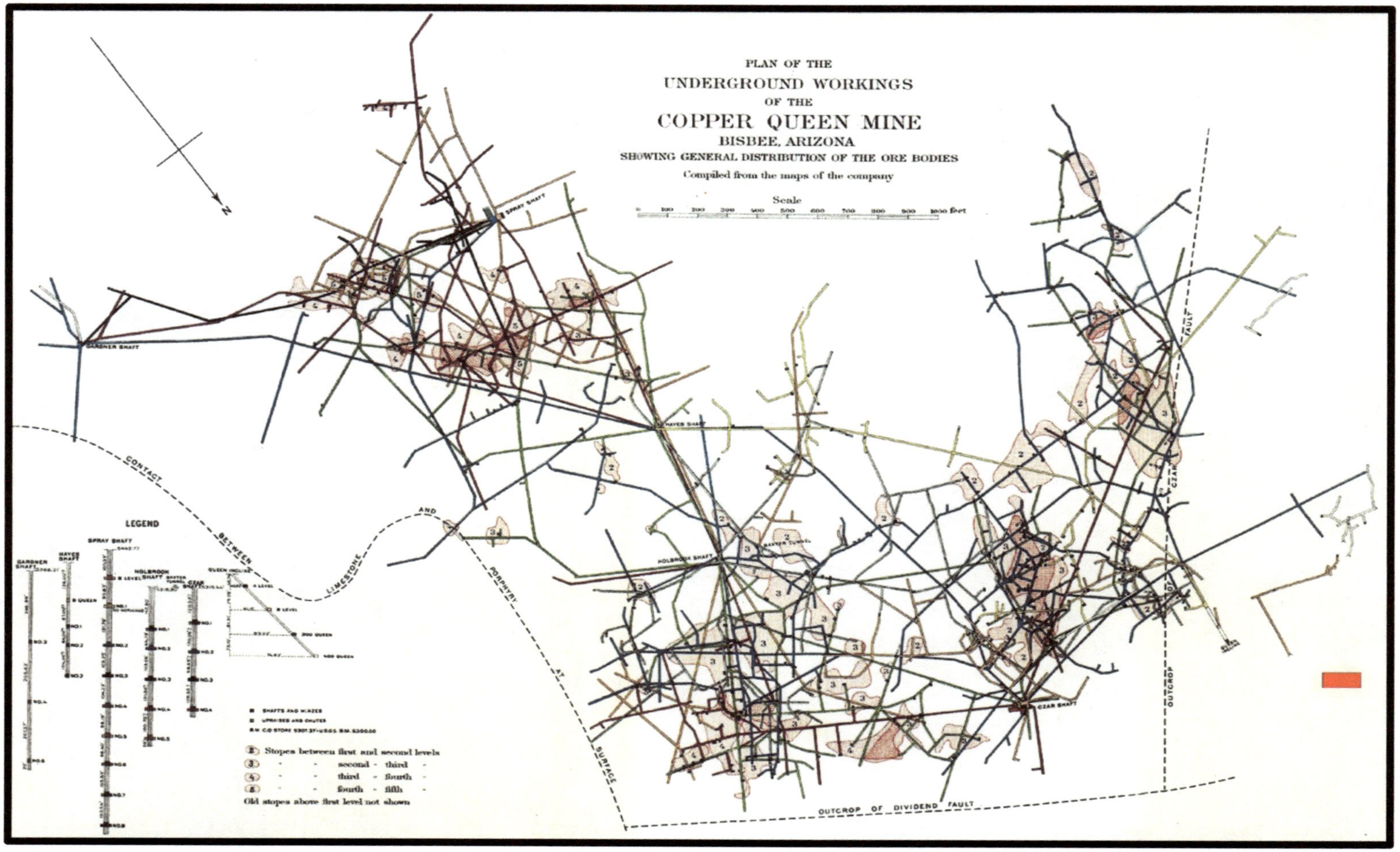

Figure 1.3.9: Map of the underground workings of the Copper Queen group of mines in 1902. Note the many miles of workings on multiple levels. The orebodies mined to this point are in red and the levels between which they occurred indicated. The vast majority of the ores mined, which were a bit removed from the Dividend fault and porphyry contact, were associated with oxidation caves. It is worth noting that as one moves east from the Czar fault the ores become increasingly deep. For reference, a red block has been inserted in the lower right corner indicating the approximate location of the Bisbee post office. After Ransome, (1904a).

Figure 1.3.10: Bisbee in 1902 and the Copper Queen Consolidated Mining Company mines studied by Ransome.

In a short summary of his work at Bisbee published in the *Transactions of the American Institute of Mining Engineers,* in 1904, Ransome noted the occurrence of malachite and azurite stalactites with the following:

> *"In the upper levels the malachite and azurite occurred as beautiful incrustations and stalactites lining caves in the limestone"* (Ransome, 1904b).

Also in 1904, the Bisbee Daily Review, commissioned a special *"World's Fair Edition,"* really a 96 page, large format information book published, not in Bisbee, but in St. Louis, which was distributed not in Bisbee, but rather at the St. Louis World's Fair of that year. The stated intent of this edition was to further Arizona's quest for statehood by noting within it was to:

> *"... refute the statements often made in Congress that Arizona was not ready for Statehood, single and alone..."*

The whole of the Territory of Arizona and its abundant resources were described in detail, including Bisbee with a couple of notes on its caves. One was a repeat of what F. L. Ransome

(1904b) written, as previously presented, but the second reference was different and interesting. When discussing the Copper Queen as "Arizona's Greatest Copper Mine," it stated that:

> *"There are some extraordinarily beautiful caves and openings in the heart of the mine, some of them very large, lined with calcite and crystals that reflect the lights of the miner's candles and form a dazzling effect."*

The *"Copper Handbook"* was a rather staid, investor oriented publication, which listed all of the copper mines in the world, including a brief description of the mine; its equipment; and often the capacity of its management or lack of, much to the chagrin of those who mismanaged mines. The 1905 edition carried a few words about Bisbee's oxidation caves under a lengthily discussion of the Copper Queen Consolidated Mining Company:

> *"The mines show numerous beautiful caves lined with calcite, some of these being of considerable size and frequently found in close association with good ore bodies"* (Stevens, 1905).

Surprisingly, the local Bisbee papers carried very little concerning these extraordinary caves, the brief note in 1904 World's Fair Edition notwithstanding, as this issue was not intended for local consumption. Perhaps, to the residents of this frontier town, these caves were commonplace and, as such, little note was taken of them. This was surely the case with the many wonderful minerals removed from the mines. As they were on display everywhere in town, yet little was ever written about these mineral specimens.

One interesting exception to this lack of locally published information was a short article in the January 13, 1906 edition of the *Bisbee Daily Review* where an unusual occurrence of fossils was noted with the below:

"FIND OF FOSSILS IN THE HOLBROOK MINE
Discovery of importance to the geological department

> *In a new cave recently entered at the Holbrook there have been found fossils imbedded in decomposed limestone of a remarkable light character. Discovery of the fossils is important to the geological department, in as much as from perfect specimens it is possible to gather information in determining the age of the formation in which they are found.*

> *Fossils have been sufficiently rare of discovery in the camp to attract attention upon their occurrence. The interest manifested is not confined to the geological department but is also participated in by the miners.*

The recent discovery was made on the 200 level of the Holbrook. The fossils come in decomposed limestone attached to the walls. This is largely impregnated with kaolinite and besides being lighter than cork is exceedingly greasy to the touch.

The fossils contained are small. No perfect specimens have yet been secured, but search is being made for them. Some that were near perfect were secured yesterday. The most have disintegrated or been broken, shells alone remaining. Aside from these features the new cave is very similar to others that have been opened in the Copper Queen mines [sic]. *"*

Marine fossils are abundant in several of the limestone horizons at Bisbee as documented so well by Ransome, (1904a) and commonly collected by many geologists (see Figure 1.3.11). The authors have seen such fossils in several caves and consider this type of occurrence neither rare nor unusual. More interesting is the occurrence of siliceous fossils associated with malachite and azurite such as those given the American Museum by Dr. Douglas (American Museum, 1918).

In a few rare instances, fossils actually replaced by malachite and/or azurite were found (see Figure 1.3.12), these are, in our view, noteworthy, but seemingly ignored. Why this rather ordinary find of fossils as described above, may have been worthy of special note is a bit puzzling.

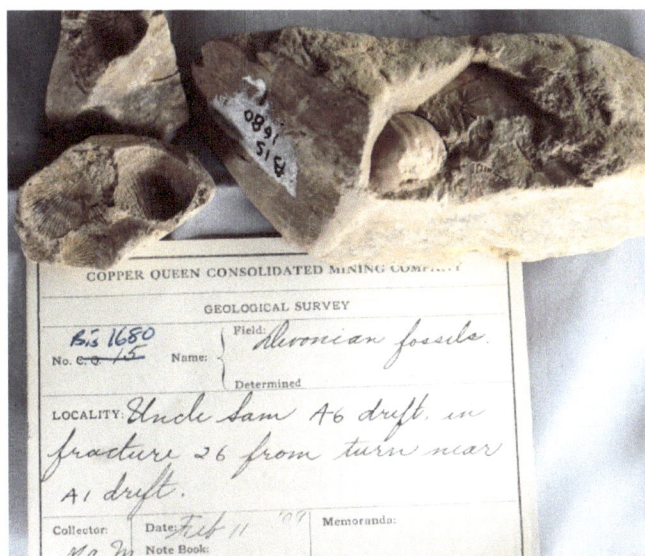

Figure 1.3.11: Devonian marine fossils from the "A" level of the Uncle Sam Mine, collected in 1909 by Copper Queen geologist for donation to Harvard College as a part of the rock and ore suite. Card – 4 inches, Harvard Mineral Museum specimens.

Figure 1.3.12: Left, malachite as a replacement of calcite fossil crinoid stem pieces on and in a mixture of goethite with manganese oxides, found in the Holbrook Mine in about 1905. View - 1.25 inches.

Also during 1906, two young women visited friends in Bisbee. In the company of a friend, Bill Gidley, they went underground in the Holbrook Mine. While they did not visit a cave during this underground trip, they were treated to a story about a cave by Mr. Gidley that they recorded.

"I am told that there is a beautiful crystal cave in the mine. The Freemasons on several occasions held meetings in it, but visitors defaced it so much by breaking off large pieces of crystal, that the company decided to fill it in with stone and waste material from blasting. The door is kept locked, and no one is admitted.

It will be interesting to know that several months later the workmen broke into a new cave, or a continuation of the one they are filling up. Mr. Gidley, not being very busy that night, conceived the idea to explore the cave. He writes it was the greatest sight he ever saw. The different specimens of lime crystallization that hung on these walls were simply grand, but to describe it is beyond power of mortal tongue. He just set amazed, and thought after seeing this magnificent display of Nature's underground curiosity shop, there remained no more on earth for mortal eyes.

The entrance is by a small hole two feet in diameter. He had to crawl on his stomach twenty feet. It was a perilous task. Then the cave took a downward course. He returned for a rope and securely fastened one end above and the other around his body. He swung out and down, over the huge cliff or rocks, about forty feet below. How large the cave is, they do not know, and perhaps never will, but he has the honor of being the first human being that ever set foot in it.

He succeeded in getting some very fine specimens of crystal, which he values. There are many specimen hunters, but the company will not allow them in the cave"
(Cole, 1908).

Bill Gidley was still working at the Copper Queen Branch in 1960 when he glowingly recounted the adventure of collecting in this cave in a bit more detail to Richard Graeme III. This cave was located just below the 300 level, between the Czar and Holbrook mines and remained open for some time before it was completely backfilled to prevent collapse due to the mining of the ore below. As backfilling took place, Bill would periodically return to the cave to collect more specimens as he could stand on the newly placed gob (backfill) and reach specimens that were not accessible during his previous visits.

Many of the specimens Gidley collected at this time were still in his possession in the mid-1960s and included some exceptional copper tinted flos ferri aragonite, (a variety of aragonite, occurring in delicate coralloidal forms). Only a few of the other pieces were copper tinted, but all were lovely nonetheless.

Robert Brinsmade, writing on *Copper Mining at Bisbee, Arizona* in the February 1907 issue of *Mines and Minerals* briefly noted that:

> *"Malachite and azurite are minor constituents as metal yielders, but some beautiful crystals, of the former especially, have been found in caves."*

In 1907, a small oxidation cave was found just above the 1050 level of the Irish Mag Mine that had formed in goethite. Stalactites of goethite in the cave were partially to completely overgrown with small octahedral cuprite crystals. In some instances, these had a drusy appearance from the myriad of tiny, sparkling crystals.

Walter Harvey Weed in his classic tome *"Copper Mines of the World"* published in 1908 noted:

> *"The limestone near the surface has caves of irregular shape, one of them 300 ft. wide and 700 ft. long, and lined with malachite, and azurite, and stalactites of calcite. These "cave" ores for which Bisbee was so long noted are now exhausted."*

An important ore strike was made in a cave located on the 550 level of the Irish Mag Mine in 1909. This was reported to the shareholders of the Calumet and Arizona Mining Company in its annual report for that year with the below:

> *"Recently we have cut a station on the 450-foot level and are drifting to cut this ore on this level. In this ore body, just above the 550-foot level, we encountered a large natural cave over 100 feet long and in places over 50 feet wide and from 50 to 70 feet in height. The north wall of this cave seems to be almost entirely oxide ore, but our development in this region has been rather slow owing to the fact that the ground is treacherous and our work must necessarily be done with caution"*
> (Calumet and Arizona, 1910).

The ore cut on the 550 level was high grade with large amounts of coarsely crystalline malachite. The ore zone extended for 30 feet into the cave from the limestone wall. A fine example of the malachite from this locality is preserved in the collection of University of Arizona Mineral Museum. This was to be the eastern most occurrence of oxidation caves in the Bisbee mines.

Chapter Four: The 1910s

Early in 1910, two references of caves with ore at Bisbee appeared in the same issue of *The Engineering and Mining Journal* (1910a). In the first report, there was a brief note under a summary of activities at the Copper Queen, where a cave in the Uncle Sam Mine was mentioned:

> *"At the Uncle Sam, the raise from the N level has encountered some good oxide ore. This raise is coming up under an extensive cave on the level above, in which was some good oxide ore."*

The Uncle Sam Mine was south of the Czar and more than 500 feet higher in elevation. The "N" level was equal to the 5th level Southwest at an average elevation of 5,500 feet above sea level. The cave had been hit by development work on the "M" level, equal to the 6th level Southwest at 5,600 feet above sea level.

This cave proved to be particularly interesting because the sand and gravel in the cave bottom contained small flakes and rounded particles of native gold. The cave was adjacent to and partially within one of the several, large silica breccias in the mine, which, on occasion, were found to contain gold. The silica breccia was undoubtedly the source of this gold. Wisser, (1927) notes a similar occurrence of gold in a cave, but does not give the location. It is possible that he is referring to this very same discovery, though it is described somewhat differently.

The second report of a cave appeared in the February 19, 1910 edition of *The Engineering and Mining Journal* was under a brief resume of the activities of the Calumet and Arizona Mining Company and was more or less a summary of what would appear in the company annual report and which is noted above. The report said that:

> *"The 450 level has been started at the Irish Mag shaft* [sic]*. A drift is to be started on this level to prospect ore found in a cave on the 550 level."*

Work by Ransome (1904a) had suggested that ore might be found in Queen Hill, the limestone knob, above and behind the Czar Mine. Except for the original Copper Queen Orebody and the small deposit exploited by the Arizona Prince Copper Company in the very early 1880s, nothing else had been found in this hill. A more determined effort to find ore in Queen Hill by the Copper Queen in 1908 was to be successful.

From workings on the 100 level of the Czar Mine and well underneath Queen Hill, prospecting upward began. Soon, several small orebodies were found and mined. This modest discovery encouraged additional exploration work that ultimately led to the discovery of the New Southwest Orebody, one of the largest orebodies ever discovered at

Bisbee. Also, a number of other smaller, but important orebodies were to be discovered later.

Figure 1.4.1: Queen Hill in 1909. The Southwest Mine exploited Queen Hill, mining rich ores from above the Open Cut of the Copper Queen Mine (center) to the Higgins Mine, seen here in the upper right. The dark area in the upper center of the photo is a silica breccia outcrop adjacent to the huge, New Southwest Orebody, which hosted several important caves.

This area of the newly discovered ore, now named the "Southwest Mine," was an important source of largely oxide ores over the next 25 years. With these near surface, oxide ores were a number of oxidation caves. The account of one cave from the *American Museum Journal* published in December 1911is presented below:

"NEWLY DISCOVERED CAVERN IN THE COPPER QUEEN MINE

By Edmond Otis Hovey

The great Copper Queen mine [sic] at Bisbee, Arizona, is most famous for the millions of tons of high—grade copper ore which have been taken from it, but it is likewise well known for the beautiful, though, small caves that have been encountered in it from time to time in the course of the regular mining operations. These caves have for the most part been found in the limestone of Queen Hill,

the eminence that forms the southwest wall of Tombstone Canon at Bisbee. One of the caverns broken into during the active life of the old Queen incline, almost in the heart of the city, twenty or twenty—five years ago furnished the wonderful green and white curved and ordinary stalactites and the stalagmites that adorn the Gem and Mineral Halls of the Museum.

There is therefore small cause for wonder that I was much interested in the report of the finding of this new cave. The word, reached my ears immediately on my arrival at Bisbee, where I had gone with three men to collect the data needed in the construction of the great Copper Queen model which is being made for the Museum through the generosity of a friend of the institution. The cave had been discovered some mouths before, but immediate steps having been taken to control access to it, its rooms and their formations were still in their pristine perfection and beauty.

Having donned regulation mine costumes early one morning, we started for the underground cavern. After descending the Czar shaft [sic] two hundred feet to the 'second level' we walked southwestward toward a point almost directly beneath the summit of Queen Hill. A quarter of a mile or more — it seemed at least a mile — from the big shaft we came to the foot of a 'raise,' up which we were drawn four hundred feet by an electric hoist. The journey from the shaft along the level through solid limestone had been cool and comfortable, but as we went up the raise both the moisture and temperature of the air increased because we had entered the 'leached ground' where the oxidation of the original ores produced heat, just as does burning coal. A few yards from the raise we reached the top of a 'manhole' cut through the heating ore. Now it was necessary to climb forty feet down vertical ladders to the heavy plank door that guarded the cave.

Squeezing through a small hole beyond the doorway, we found ourselves at the bottom of the cave in a small room whose ceiling scarcely permitted one to stand erect. The bright light of our acetylene mine lamps showed that the room was lined with alabaster, tinted a delicate green with carbonate of copper. Walls and ceiling were comparatively smooth but incrusted with minute crystalline surfaces that glittered in the rays from our lamps, while the floor was uneven with knobby clusters of calcite and held here and there a shallow pool of limpid water. The upper exit from this first chamber was almost closed with great blocks of rock that fell from the ceiling so long ago as to have received their own coating of dripstone. Worming our way upward among these for a few yards, we emerged into a clear chamber fully thirty feet high and forty feet across. The floor rose at a steep angle and its coating became part of the base of a great stalagmite.

Figure 1.4.2: Illustration of the cave from Hovey, (1911) and captioned, as follows: "*The large room (thirty feet high and forty feet across) of the cave was one of the most beautiful sights imaginable in the brilliant illumination of our acetylene mine lamps. Its chief feature was the great greenish white stalagmite (fourteen feet high) rising at its upper end, so impressive in size and setting, so beautiful in outline, ornamentation and surroundings that it seemed little short of vandalism to destroy or mar it, or any part of the cave which it adorned, although in the interests of science.*"

This large room was one of the most beautiful sights imaginable in the brilliant illumination of burning manganese ribbon. Its floor was a thick mass of dripstone, its walls were partly smooth white calcite and partly, toward the top, the deep velvety brown, red and black of the iron— and manganese—stained residue of the decomposed country limestone, while the ceiling was mainly of the limestone but banded with sheets and small stalactites of calcite. These occurred along the old cracks in the mountain mass, which formed the channels for percolating waters, an important factor in the formation and incrustation of the cave. The lower part of the walls was thickly covered with botryoidal clusters of white calcite, some areas of which were tinted a delicate salmon color with carbonate of manganese.

The chief feature of the room was the great greenish white stalagmite rising at its upper end and reaching almost to the ceiling. So impressive in size and setting, so beautiful in outline, ornamentation and surroundings was this wonderful object it seemed to us little short of vandalism to destroy of mar it, or any part of the cave which it adorned, although in the interests of science or the necessities of mine operation. This stalagmite is about fourteen feet high above the shelf of limestone on which it stands and its diameter at the same point may be taken as being fourteen or fifteen feet. Three feet above the shelf the column is ten feet through. Stalagmite is of extremely slow growth and even under the more favorable conditions prevailing at Luray Cave, Virginia, where measurements have been made, such a mass would have required more than 67,000 years to form; hence. it is safe to assume that this cavity in the Queen Hill has had its present size and shape for a much longer period than that, since the rainfall is less and the consequent solution slower in Arizona than in Virginia, though evaporation and consequent deposition are conversely more rapid in Arizona. The stalactite growth above this stalagmite was insignificant.

Climbing up the congealed waterfall forming a smooth apron in front of and below the stalagmite, we passed to the left of the column over a floor carpeted with coarse botryoidal clusters of calcite and clambered through an opening in the black rock into a room that might be considered the fourth story of the cave. Immediately at our right was a compound stalactite which our miner associates promptly called the elephant's ear, while a few feet beyond was a remarkable stalagmite three feet in diameter and rather more than three feet high, which with its smaller stalactite and its accompanying crystal-covered floor and wall formed a charming grotto.

This stalagmite was noteworthy on account of the radiating clusters of pointed calcite thickly set all over it but diminishing in size from the bottom of the column upward. It has been commonly held that such crystals could be formed only under water, but conditions here indicate that there has been no submergence or filling of the cave since it was formed and we must conclude that in a region of extremely rapid evaporation crystals will grow from a solution flowing over a surface.

The upper wall of this room was formed by a great block of fallen rock which has received the drippings of a lime-bearing watercourse. Stalagmite was formed on its top, while ribs of calcite, some of which were complete lines of crystal tufts, projected close together from its sides. Narrow, drip— stone—lined passages on either side of this block led to a series of three small rooms one above another, the last of which was so low that an adult could hardly squeeze his way into it.

These upper rooms were characterized by abundant stalactites and practically no stalagmites, contrasting with the conditions in the lower rooms where the stalagmites predominate at the expense of the stalactites. One of the most beautiful small features of the cave was the occurrence on the walls of one of the upper rooms of long acicular crystals of delicate green calcite grouped paintbrush fashion on small botryoidal masses of the same material. The cave extended up slopes averaging thirty degrees, through a vertical distance of about eighty feet and nowhere exceeded forty feet in width and thirty feet in height.

Inasmuch as the cave was doomed to ruin through mining, the company generously furnished the men and the means for removing at infinite pains the grotto and such other formations as we desired, for transporting them to New York. This material is now at the Museum and there will be in place and on exhibition a reproduction of this most beautiful underground chamber" (Hovey, 1911).

This apparently magnificent cave was actually in the Southwest Mine even though Hovey accessed it through the Czar Shaft. The "raise" up which he was hoisted was the first, of several, Southwest interior shafts. At the time of discovery, none of the surface access adits or tunnels for the Southwest Mine had been developed, thus the need to use the Czar Shaft to reach the area.

The location of the cave was on the 4th level of the mine, but accessed by climbing down from the 5th level. Even though the authors have spent a good deal of time in the general vicinity of this cave, absolutely no sign of its existence remains. So complete was the backfilling and collapse that, even as long ago as 60 years, there was no indication of this cave having ever existed.

The several tons of mineral specimens recovered from this cave became an important display at the American Museum of Natural History in New York. An article in *The Engineering and Mining Journal* told the story in 1916:

"Copper Queen Cave in New York
By Walter L. Beasley

In the course of mining operations at the Copper Queen Mine, Bisbee, Ariz., a beautiful stalactite-lined cavern was recently broken into. Exploration proved this to be a series of grottos which contained a wealth of beautifully shaped and copper-tinted formations.

Fortunately, for the preservation of such beauty to posterity, news of the discovery reached Dr. Edmund Otis Hovey, curator of geology at the New York Museum of Natural History. Before mining necessitated the destruction of the caverns, Dr. Hovey and three assistants visited Bisbee to attempt the preservation of at least a portion of the grotto formations.

In the interest of science the Copper Queen Company placed the caverns at the disposal of Dr. Hovey for the museum. The grottoes had an age estimated at 25,000 years based on the length of time required to form the slow-building stalactites. The chief feature of beauty of one of these grottoes was a great greenish-white stalagmite rising from the limestone floor nearly to the ceiling, a height of about 14 ft., its diameter at the base being about 15 ft.

Stalagmite is of extremely slow growth, and even under the more favorable conditions of growth at Luray Cave, Virginia, where actual measurements have been made, such a 14-foot mass would have required more than 17,000 years to form. Hence it is safe to assume that this grotto in Queen Hill is

Figure 1.4.3: From Beasley, (1916), labeled as: *"Reproduced Copper Queen grotto, New York Museum of Natural History."*

much older than 17,000 years, since the rainfall and consequent solution is considerably slower in Arizona than in Virginia, through evaporation and consequent deposition is more rapid.

Having decided to attempt the reproduction of a portion of the caverns for exhibition purposes at the New York museum [sic], Dr. Hovey and his assistants spent many weeks in skilled and patient work in carefully dislodging their required specimens. Some idea of the care required is obtained from the statement that the stalactites and stalagmites weighed from one to over 800 lbs. each. Fifty boxes of the beautifully tinted blue and green formations were collected from the ceilings, walls and floor of the grottoes and shipped to New York.

So beautiful were the caverns in their outlines and tinted formations that the curators were continually impressed that their work was little short of vandalism in the name of science. The accompanying illustrations of the New York reproduction of the cavern, however, will show why they ceased to harbor regrets.

Figure 1.4.4: From Beasley, (1916), labeled as: *"Corner of reproduced Copper Queen grotto."*

This work of assembling and jointing the large number of stalactites to form this museum cavern reproduction was a feat successfully accomplished under the direction of William A. Peters, an artist of the museum staff who personally accompanied the expedition to Arizona.

The final reproduction of the cavern at the New York Museum of Natural History occupies a space 14 ft. high by ten ft. long. The outside of the cave is roughly finished in limestone blocks and the interior portions of which are shown in the accompanying illustrations, is viewed through a large glass panel. The natural color beauty of certain of these stalactites has been heightened by the use of concealed electric lights placed in many sections of the cave" (Beasley, 1916).

This popular display remained in the American Museum until it was removed for remodeling in the mid-1970s. Only one impressively colored stalagmite remains on display (see Figure 2.7.25). Most of the other copper tinted formations from this display were exchanged by the Museum. Subsequently, a few of these specimens ended up on the market and were quickly purchased by collectors and by other museums. Several of the larger, more impressive, stalactites from the display at the American Museum were incorporated into the oxidation cave display in the Bisbee Mining and Historical Museum in 2006 and can be seen in Figure F-1. This is a most fitting destination for some of the specimens, as they have returned home to Bisbee, and continue to be appreciated and serve as a reminder of just how magnificent these caves were.

Figure 1.4.5: Right, specimen of copper tinted, flos ferri aragonite from the Southwest Mine, which was formally a part of the Copper Queen Cave display in the American Museum of Natural History. Specimen - 15 inches.

Walter Harvey Weed, a consulting geologist who was well on his way to becoming the foremost expert on copper mining in the United States at the time, made the following reference to Bisbee's caves in the August 12, 1912 edition of *Mines and Methods:*

> *"The great crystal studded caves, for which the district was noted in the early days of its workings, are now worked out, the stopes extending deeper to the sulphide zone in all the older mines, though a new cave was opened up two years ago with mossy malachite walls, and a chamber 75 ft. or more across. No other district in America has furnished the amount or variety of malachite given by the Copper Queen, nor handsomer crystals of azurite, cuprite and chalcotrichite."*

In early 1913, the largest cave ever found at Bisbee was hit in the Shattuck Mine. Because the discovery was so unusual, the story of this find was widely circulated. It was first printed in the magazine *Science* and later in several national newspapers. Below is the original article from the March 12, 1915 issue of *Science* where the receipt of specimens from the cave was also reported:

> *The Michigan College of Mines has received a collection of minerals from the Shattuck Cave, near Bisbee, Arizona, one of the wonders of the mining world. This cave was opened in 1913 by a drift on the third level of the Shattuck Mine. When the miner who had been drifting in this part of the level returned one night after a heavy blast, he found that the working face had entirely disappeared and that before him was a great opening reaching farther than his light would shine.*

> *Looking upward he could see tiny lights flashing and believing that they were stars he ran back to the shaft, declaring that he had blasted a hole clear through to surface. Mine officials investigated at once and found that a great natural cavern had been opened up, circular in shape, 340 feet in diameter and 175 feet high. It*

was a virtual fairyland of beauty, myriads of crystals in the roof reflecting back the lights from the miners' lamps. Walls, roof and floor were covered with great clusters of crystals, and near the center of the cavern a cluster of stalactites hung from the ceiling in the form of a great chandelier 40 feet long. The crystals were for the most part pure white, but in places where the filtering waters had contained iron and copper, the beauty was enhanced by great transparent stalactites and stalagmites, some ruby red, others a clear emerald green or -azure blue. The mining company illuminated the cave with electricity and has allowed thousands of visitors the privilege of seeing it.

An attempt was made to have the Smithsonian Institution at Washington remove and reproduce a portion of the cave, but nothing came of it. It is because the mine operators have now found it necessary to fill the cave with waste rock that the Shattuck-Arizona Mining Company sent the specimens to the College of Mines. Superintendent Arthur Houle, of the Shattuck Company, is a brother of Professor A. J. Houle of the college.

The Shattuck Cave was spectacular by any measure. The Shattuck and Arizona Copper Company realizing this cave was a treasure, quickly took steps to protect it from the ever-present and opportunistic mineral collecting miners. Electric lighting was installed and hundreds of visitors were allowed to visit the cave, but under strict, escorted supervision, to preserve the fragile formations.

The Shattuck Cave was undoubtedly the most visited and most photographed cave at Bisbee. A local photographer even produced a series of photographs, which he offered for sale. Examples are shown in Figures 1.4.7 and 1.4.8.

Perhaps the most succinct description of this cave was written by the noted Arizona geologist Philip D. Wilson in the April 11, 1914 of *The Engineering and Mining Journal*. Wilson described the cave as follows:

"A Cavern in the Shattuck Mine

Synopsis – Large natural caverns have been found in the mines of the Bisbee district in Arizona, where limestone is associated with porphyry. The most recent cavern to be opened is that is in the Shattuck Mine, the largest and most beautiful so far found.

¤

Recently in the course of development work in the mine of the Shattuck Arizona Copper Co. at Bisbee, Ariz. there was encountered what is, both on account of its size and rare beauty, probably the most remarkable natural cavern in the country; perhaps in the world. Other caverns somewhat similar have been opened in the

Bisbee district. A cavern in one of the older portions of the Copper Queen mine [sic] has become famous locally as the auditorium in which a large fraternal organization met in solemn conclave in the earlier years of the camp, and another cavern brought to light more recently in another part of the same mine has been transported almost in its entirety to the Museum of Natural History in New York. But both of these and others of less renown are insignificant before the one more recently opened.

It was first discovered by a drift on the 300-ft. level which fortuitously struck it in its lowest and in a central point. A drift a few feet on either side would have passed beneath it and have left it perhaps unknown for years. In shape it is a huge lens approximately following the bedding planes of the inclosing limestone at an inclination of about 35^0 and it is roughly circular in horizontal projection. Its upper extremity is 172 ft. above the 300-ft. level and the diameter of the circular projection is 340 ft. The vertical distance between roof and the floor where its height is greatest has been roughly estimated at 80 feet.

One's first impression of this great cavern now electric lighted, with its stalactite studded dome, is that of a shadowy interior of a Gothic cathedral. Close examination reveals myriad forms of calcite, crystalline and amorphous, with all its vagaries of structure and color. It is apparent from the structure that a lime-impregnated solution has filled portions of the cavern subsequent to the original formation of the stalactites and stalagmites; left its quota of mineral as arborescent, coral-like deposits on the stalagmites, and afterward drained away. In many cases a second generation of stalagmites has formed, and in places there is evidence that this alteration of aerial and subaqueous deposition has taken place several times. A unique occurrence is shown in the first that at the left, of the accompanying photographs [Figure 1.4.6]. Known in local mine parlance as "calcite wiggletails" these curious serpentine growths ranging from 1/8 to ½ in. in diameter, emanated from the limestone hanging wall in the most amazing spirals and volutes or shoot out at every conceivable angle. Each one, as described by Prof. Alexander H. Phillips, of Princeton University, seems to be a complex parallel growth of elongated and curved rhombohedrons.

Figure 1.4.6: Photographs taken of the Shattuck Cave in 1913, note the man in the center of the upper right photo, (Wilson, 1914).

and below that again is an irregular mass of siliceous breccia. The breccia zone extends to within a few feet of the 700-ft. level, where it rest on a sill of granite porphyry of great lateral extent and variable thickness. Throughout the detritus zone and the mass of siliceous breccia are scattered shoots of high-grade copper and lead-silver ore.

Associated with this breccia are found several rare minerals unique to the Shattuck, most notably of which is a deposit of a rare copper-lead vanadate, a non-zinciferous cuprodescloizite.

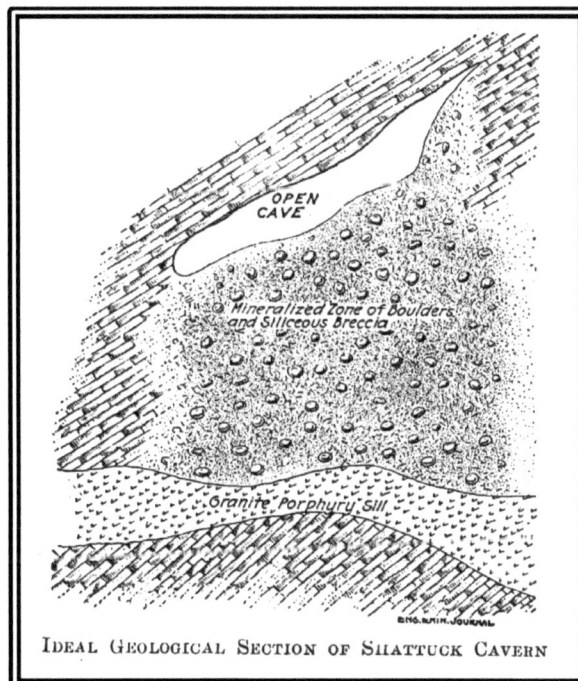

Figure 1.4.7: Idealized cross section through the Shattuck Cave, (Wilson, 1914).

64

Several conjectures have been advanced to explain the origin of this cavern. It is probable that the shrinkage contingent upon the solidification and cooling of the intrusive rock mass shattered and opened the rock mass for a great distance above it. This left a large open space easily accessible to the acid meteoric waters which enlarged the cavity to its present size and left it ready for the calcium carbonate" (Wilson, 1914).

Wilson's closing words regarding the calcium carbonate are a reference to the abundant and magnificent calcite and aragonite formations so vividly on display in the cave. The variety of forms was impressive, even at the time when such caves were still being found. As the accompanying photographs demonstrate, the cave was well appreciated.

Figure 1.4.8: Calcite stalactites and anthodites, Shattuck Cave 1913. Dimensions unknown.

Figure 1.4.9: A trio of photographs taken of the Shattuck Cave during March 1913. Dimensions not recorded.

Wilson's novel idea that the opening was created by the shrinkage following cooling of the intrusive was, in our opinion, erroneous. Indeed, his cross section, as shown Figure 1.4.7, clearly leads us to this conclusion. Most significantly, the subsequent mining of the mineralized zone below the cave showed that it was formed through the same processes as the other oxidation caves in the Shattuck and other mines at Bisbee. Namely, the oxidation of sulfides and the subsequent reduction in volume of the remaining oxides caused the opening to form.

The only difference between this cave and many others was the presence of the granite sill reasonably close below. This same granite unit was directly associated with many of the orebodies found in the Shattuck and other mines in this general area. No other similar shrinkage features as suggested by Wilson, were found along this extensive sill/dike system. However, numerous silica breccia masses, such as that illustrated by Wilson in Figure 1.4.7 and associated with the porphyry and caves were noted by other geologist (Bonillas, et al., 1916, Trischka, 1928, 1932).

Figure 1.4.10: Shattuck Mine in 1910.
The cave on the 300 level of this mine was below the surface depression marked by the three limestone bluffs near the center of the photograph.

The Arizona Elks held their annual meeting in Bisbee in early April 1916. The highlight of this gathering was a luncheon and a dance held on Saturday, April 9th in the Shattuck Cave. The March 25, 1916 edition of the *Arizona Silver Belt* newspaper of Globe, Arizona reported on the arrangements:

> *"This latter will be the real unique feature of the convention, Joe Walker of the Shattuck force has undertaken all the arrangements. Luncheon will be served and the Copper Queen band will furnish the music for the dance which will be held on*

the flooring in the cave. Special illumination and lighting effects will throw out more clearly the beauties of this crystal palace, of which nature has been the architect."

Ultimately, the Shattuck Cave, like most others, was mined. Large amounts of silver-rich, lead carbonate (cerussite) ore was mined from within one part of the cave, and copper ore mined from below the other part of the cave. Some backfill was placed in the cave, but only a small part was filled. This was because of a lack of waste rock above the 200 level to use as backfill. Before long, the cave partially collapsed. Soon, house-sized boulders had dropped into the opening forming a jumbled mass. Perhaps a third of the cave remained open when the authors last visited it in the early 1990s and it was much less impressive than described. In no small part, this was because the Shattuck Arizona Mining Company thoughtfully salvaged as many of the formations as possible before starting the mining and filling, the cave.

Numerous museums received the materials recovered from the Shattuck Cave. In 1914, the Harvard Museum noted it had received some material by writing:

> *"The unique collection of stalagmites from the Shattuck Mine at Bisbee, Arizona deserves special mention. The Company gave the stalagmites, and the Museum paid the necessary expenses. Three of the stalagmites average six feet high with base rock enough to permit a natural group. There is a claw like stalagmite four feet high showing reddish coloration on the upper side and snow white on the under side. A large number of smaller stalagmites of remarkable shapes are colored yellow, cream, purple, red and pink. Some of them are of the most pure white. Some have coral-like forms, and others are botryoidal or branching"* (Harvard College, 1915).

By 1917, these Harvard College specimens had been incorporated into a reconstructed cave display as reported by Sayles, (1917) in his summary of the geological collection:

> *"The installation of the unique collection of cave deposits from Bisbee, Arizona, in the new hall case, was successfully accomplished. The dark hall and artificial lighting give a realistic effect not attainable by sunlight."*

The systematic recovery of materials from this huge cave continued for some years and many other museums received specimens as noted in the *Lincoln* (Nebraska) *Sunday Star* on April 9, 1922:

> *"On Wednesday of last week, the Malben geological collection in the university museum received an important addition from southern Arizona, consisting of three large cases of aragonite, calcite stalactites and stalagmites and associated rocks and ores from a deep cave near Bisbee.*

This aragonite cave was discovered 1600 feet underground and completely despoiled. The material recovered is being shipped to the universities of California, Texas and Nebraska, and also to the American Museum of Natural History, New York. The largest collection aside of that secured by the University of Arizona was received by Nebraska..."

This cave was also of mineralogical interest. In addition to the cerussite mined from this cave, the authors have found anglesite (lead sulfate); galena (lead sulfide), bromargyrite (silver bromide), murdochite (copper lead chloro/bromo oxide) and plattnerite (lead oxide). All occurred in minor amounts intermixed with the few spots of remaining ores. There is every reason to believe that most of these mineral species were distributed throughout the ores and relatively abundant.

Mineral specimens from the Shattuck and many other caves were collected by many Bisbee residents, as were the other impressive minerals from the mines. In more than a few instances, these specimens were placed on public display. Private collections were in banks, hotels, businesses and, of course, in many of the saloons, for all to enjoy. Even the YWCA contained a mineral display. At least one local jewelry store used cave specimens as a backdrop for its wares n the show window as illustrated in Figure 1.4.11.

Figure 1.4.11: Several, large cave mineral specimens in the show window of a Bisbee jewelry store, C-1915.

A second, geologically important cave was discovered in the Shattuck Mine about the same time. This opening was called the "Shattuck Vanadium Cave" and was a most unusual discovery (Wells, 1913). Vanadium minerals lined the cave, which Wells (1913) identified as "cuprodescloizite." Much later, other scientist studying the material, showed the principal mineral found in the cave to be actually mottramite, a lead copper vanadinate hydroxide (Tabor and Schiller, 1930) and not cuprodescloizite as reported by Wells (1913).

The Shattuck Vanadium Cave was roughly 60-foot in diameter by 15 feet high and developed as an oxidation style cave over a mixed lead/copper orebody. It formed largely within a mass of soft, mixed manganese/iron oxides. Stalactites of mottramite collected from the cave at the time of discovery were several centimeters in length and up to 8 mm in diameter with a radiating, internal structure were found here. Wells (1913) wrote the below concerning the stalactites:

"The stalactitic form of the mineral indicates that it crystallized by the evaporation of downward migrating solutions and its chemical character shows that the solutions were products of oxidation."

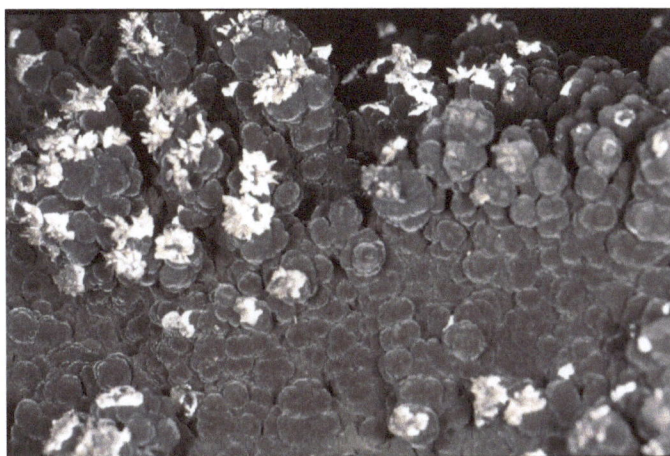

Figure 1.4.12: Brown-black mottramite with minor calcite, 8 stope, 600 level, Shattuck Mine, view – seven inches.

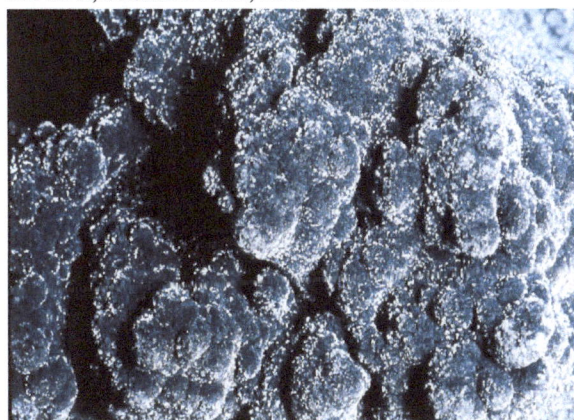

Figure 1.4.13: Olive-brown mottramite, 8 stope, 600 level, Shattuck Mine, view – three inches.

Figure 1.4.14: Drusy crystalline, black mottramite, 8 stope 600 level Shattuck Mine – 3.2 inches

Botryoidal forms up to ¾ inch high occurred in the upper portions of the opening, where the walls were covered with dark olive-brown to brown-black, velvety, reniform masses as shown in Figures 1.4.12 and 1.4.13. The lower portions of the cave were littered with angular fragments of altered rock and goethite that were covered with a 0.1 inch, drusy crust of mottramite (see Figure 1.4.14). This form indicated a subaqueous depositional environment in this, the bottom part of the cave. Other vanadium minerals recognized from this cave are descloizite, a lead zinc vanadinate hydroxide as well as tiny amounts of the very rare species, volborthite, which is a hydrated copper vanadate hydroxide.

The cave and surrounding area were mined for the vanadium ores and was called "8 stope," because it was the eighth stope developed on the 600 level. In spite of the local abundance and richness of these vanadium ores, no other deposit of vanadium ore has been discovered in Bisbee.

In 1914, the well-known economic geologists, Bateman and Murdoch, were hired by the Copper Queen to do an investigation on the supergene enrichment at Bisbee. While involved in this study, they had a number of discussions with the miners who, were well noted for their observation skills. Miners told them of a place where malachite flakes had been deposited on stalactites after the cave had been opened by mining. Intrigued by this possibility, Bateman and Murdoch went to the cave located on the 4th level of the Southwest Mine to investigate.

They did find flake-like pieces of malachite on more than a few of the stalactites. Bateman and Murdock proposed that the source of the malachite was not recent, as the miners had supposed. Instead, they thought the malachite had always been there but encased in the stalactites. It was only now exposed on the surface of the cave formations because of the etching of the calcite by mild carbonic acid solutions. The acid had formed in the cave

Figure 1.4.15: Malachite on calcite stalactites exposed by the selective etching of the aragonite and calcite, 4th level, Southwest Mine, view - 34 inches.

Figure 1.4.16: Close up view of the stalactite illustrated in Figure 4.13 showing malachite on calcite stalactite exposed by the selective etching of the aragonite and calcite, 4th level, Southwest Mine, view - eight inches.

through absorption of atmospheric carbon dioxide into the water condensing on the formations. Bateman and Murdoch never noted in their report, whether or not the miners accepted this explanation.

Condensation corrosion, as it is termed, occurs when condensation equilibrates with the cave atmosphere with the absorption of increased amounts of atmospheric carbon dioxide. It becomes acidic and dissolves both the limestone bedrock and the speleothems (Auler and Smart, 2004). Condensation corrosion has been recognized in many caves worldwide and is well documented in the technical literature (Hill, 1987, Dublyansky and Dublyansky, 2000). Thus, it would not be a surprise to find this occurring in the carbon dioxide rich mine environment.

The authors visited this cave on the 4th level of the Southwest Mine several times over the years and made an effort to confirm the hypothesis of Bateman and Murdoch. For us, there was the additional; why malachite, a carbonate mineral, which like calcite typically reacts readily with acid, did not etch away as well. There was no evidence for copper mineral coloration in the calcite forming in drip areas below the corroding stalactites. This suggested copper was not mobilized, thus that the malachite was not being dissolved. Indeed, in one place, the area under the corroding stalactite was littered with malachite flakes, liberated as the stalactite dissolved.

Just as Bateman and Murdoch had suggested, the authors found it appeared that the calcite was being selectively removed by the very weakly acidic condensation. The flake-like malachite pieces remained in relief, newly exposed, on the stalactite surfaces. Furthermore, after a closer look, it became apparent to us that it was actually aragonite that was being preferentially dissolved, followed by lesser amounts of calcite. This is not surprising as aragonite is relatively more soluble than calcite in carbonic acid with increasing concentrations of atmospheric carbon dioxide (Wey, 1959). It was noted that the malachite was etched as well, but ever so slightly when compared to the aragonite and calcite surfaces of the stalactites. The malachite flakes are preserved on the surface of the deeply furrowed stalactites for two reasons. First, malachite dissolves much slower in weakly acidic cave water than either aragonite or calcite. Second, the reaction with the surrounding bulk aragonite and calcite in the stalactites may sufficiently buffer the available acidic waters before they could fully react with the malachite. In summary, a slight difference in the solubility of the three different carbonate minerals and their rates of dissolution may favor the preservation of the malachite flakes over the bulk material in the stalactites. The amount of malachite contained in these stalactites was most impressive. It seemed to have been deposited simultaneously with the calcite/aragonite as uncommonly large particles and in locally concentrated amounts. The authors have never seen a similar amount of malachite evidently co-deposited with aragonite and calcite before nor since in any of the many other caves visited and in but one of the thousands of cave specimens studied (see Figure 2.7.27).

Figure 1.4.17: The large, dark, truncated, stalactite just right of center is being corroded, liberating flakes of malachite, which have accumulated like dust on the brown flowstone below. Note the separation and cracking of the columns in the center, reflecting continuing subsidence of the cave area, 4th level, Southwest Mine, horizontal view – six feet.

A broken fragment of a large, fallen stalactite was removed for further study. The nearby photograph, Figure 1.4.18, is of a polished cross section of this stalactite with alternating zones of copper tinting. It is interesting to note that the coloration varied greatly over a short distance. This was a result of included malachite, as coarse particles, causing some areas to appear quite dark. The stalactite contained areas composed of calcite, but is largely aragonite as indicated by the fibrous appearance. This is interesting. Typically, aragonite alters to its more stable polymorph, calcite quite readily. However, in this case the aragonite had been preserved, possibly because of the presence of metal ions.

During 1917, yet another cave was discovered in the Shattuck Mine. A brief note in *The Engineering and Mining Journal* stated that:

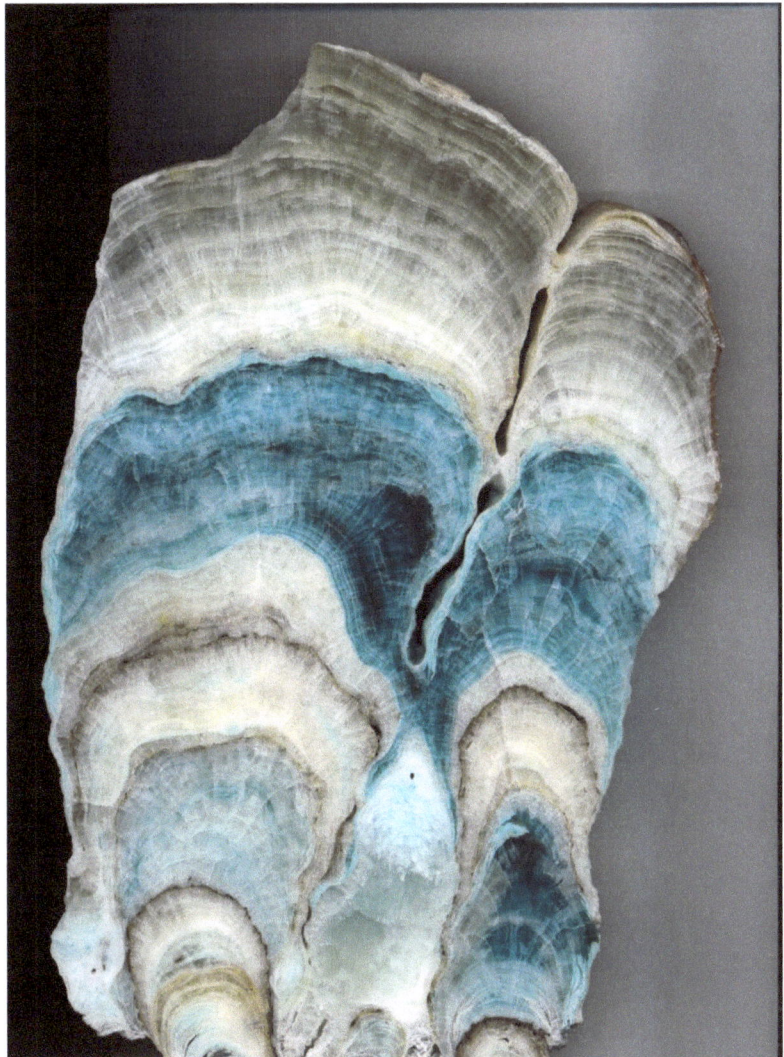

Figure 1.4.18: Polished aragonite/calcite stalactite cross section showing the episodic nature of copper coloration and mineral inclusion during deposition, 4th level, Southwest Mine, specimen – 12 inches.

"In [a] *cave on the 100-ft. level, high-grade ore lines west and north walls; some of the ore runs 40% copper; cave is 120 ft. long and about 40 ft. wide."*

Cave minerals continued to be salvaged ahead of mining when the caves were discovered and these specimens remained in demand. The advertisement below is after one in the October 1917 *American Journal of Science* and lists some recently salvaged cave minerals from one of the Copper Queen's mines. By this time, copper tinted specimens had become much less common, but they were still very much sought after and the relatively high price of $22.50 as quoted for the one piece in the ad in Figure 1.4.20 reflects a good market for such pieces. This price would have placed the specimen beyond the means of an average collector of the day.

LIST OF CHOICE SPECIMENS

Remarkable **calcite stalactites** from the famous Copper Queen Mine, at Bisbee, Arizona. Coral-like formation and rich in colors; some specimens have three and four colors in different shades of green and blue. A description can give only a faint idea of their beauty. Only a very small amount was found. They will make a wonderful centerpiece in any cabinet.

Specimen, 9" long and 5" wide with small calcite xls. attached; white, blue and green. $22.50.
Another 7½ x 4½", 4" high; beautiful shades of light and dark green. $15.
Another 5 x 4½"; 3½" high; shades from white to dark green. $8.
Another 5 x 3½"; 2½" high; shades from white to dark green. $4.

ALBERT H. PETEREIT

81-83 Fulton St., New York City

Figure 1.4.20: Advertisement for Bisbee cave minerals after *The American Journal of Science*, 1917.

To put this price in perspective, during 1917 the wages at Bisbee averaged $5.72 a day for underground workers and $4.26 a day for those men working on the surface (Mills, 1958). Also, it is worth noting that Bisbee miners were much better paid then working men in most parts of the United States at this time.

Chapter Five: The 1920s

By the 1920s, much of the mining at Bisbee had moved to the east and to even greater depths, depths where the orebodies had not been fully oxidized and caves had not formed. And too, another change took place as well which impacted the discovery of more caves. Small time lessors took over the mining of parts of the older mines, properties that the Copper Queen had operated for 40 years. These lessors typically mined small patches of known ore found along the margins of old stopes. If any exploration was undertaken by these poorly financed small miners it was modest. For most of the 1920s, few caves were discovered as little new ground was opened up. Nevertheless, in 1926, an interesting find was described in *The Engineering and Mining Journal*:

"New Cave Found in Copper Queen

Lessors working in the Southwest Division of the Copper Queen mine [sic] of Phelps Dodge Corporation at Bisbee, Ariz., recently broke into a limestone cave 220 ft. long, 80 ft. wide, and with an average height of 40 ft. Its outline is irregular and its height variable, being the highest in the center. In describing this cave, Carl Trischka, Chief Geologist for the company says:

Figure 1.5.1: The "Bath" in the Southwest Mine Cave as noted above. The men present are supervisors or engineers as indicated by the carbide lamp type as well as the boot styles, 1926.

'There are many stalagmites in the cave, and many delicate and beautiful crystals of calcite on the sides and floor. The variety of shapes due to various habits of growth calcite under the conditions of formation, is both great and delightful. The

76

bottom has the appearance of a frozen stream. Its surface is level and as smooth as ice, but it is calcite, and the illusion of a frozen stream or glacier is heightened by the cascade effect in which the flat surface terminates.

Up in one corner of the cave there is a shallow pool of water, which is surrounded on three sides by columns of stalactites and stalagmites which have grown together, forming a curtain-like effect. These columns have growing on them delicate crystals which are intergrown and others branchlike. The coloring is white to cream, and the columns when struck, give forth a bell-like sound. This is called The Bath (see Figure 1.5.1).

Figure 1.5.2: Above, central part of the cave on the 7ᵗʰ level, Southwest Mine showing the large stalagmite and "Pagoda" as described so well by Trischka (1926) below, photograph by Peter L. Kresan, 1977.

Standing almost in the center of the cave, there is a 14-ft. stalagmite, which is 5 ft across at its base, and its top and side are covered with ripple marks and small ridges which form irregular traces and patterns. A very fantastic Chinese pagoda effect is created by the calcite, which has formed on an inclined shelf of rock.'

Figure 1.5.3: Massive hematite (upper left) exposed in the ceiling of the cave on 7th level, Southwest Mine, photograph by Peter L. Kresan, 1977.

78

Figure 1.5.4: One of the authors under, mixed goethite and hematite with blocks of largely unaltered limestone in the walls of the stope below the cave described by Trischka on the 7th level, Southwest Mine.

This was a beautiful cave by any standards. It was highly decorated with spectacular stalactites and stalagmites, as so eloquently described by Trischka. Like many oxidation caves, it was partially within massive iron oxides. In this instance, it was hematite as shown in Figure 1.5.3. As was typical, the cave had developed over a copper orebody, which when mined out, left stope walls of colorful, massive hematite and goethite, as illustrated in Figure 1.5.4.

Figure 1.5.5: Right, non-miner visitors to the 7th level, Southwest Mine Cave in 1928. Note the stalactite in the hand of the man on the right. Such collecting in this cave was unfortunately common. Unfortunate here because perhaps this was the only cave in the mines at Bisbee, which could have been preserved. Because the cave was so high in the mountain, no ore was above, allowing the opening to safely remain unfilled. However, over the years many more such pieces would be removed by fascinated visitors, collectively resulting in substantial damage.

This cave was to be the west most ever found at Bisbee. It was located some 80 feet above the 7th level of the Southwest Mine at about 5,780 feet above mean sea level and near the sideline with the Higgins Mine property. Because access was much easier through the portal of the Higgins Mine, it soon became known as the "Higgins Cave" even though it was actually located within the Southwest Mine.

Early in the summer of 1954, Carl Trischka told Richard Graeme III of this wonderful cave and how to access it through the Higgins Tunnel. Carl's glowing description was temptation enough and soon after, Richard III made his first of many trips over the next 50 years into this cave.

During a visit in 1973, Richard Graeme III enlarged a small hole in the cave bottom and along one edge. This led to several small, connected rooms which had formed between the limestone and the oxides by the shrinking of the oxides.

Figure 1.5.6: Aragonite on calcite in place, 7th level, Southwest Mine, view - 17 inches.

One room contained extraordinary stalactite-like formations to several feet in length and composed totally of aragonite crystals, as shown in Figures 1.5.6 and 1.5.7. The crystals were several inches in length and intergrown in jackstraw-like fashion. Minor amounts of white dolomite, as partial overgrowths on the aragonite, was a common associated mineral.

Figure 1.5.7: Aragonite, in stalactite-like forms to 30 inches, on calcite with minor white dolomite, 7th level, Southwest Mine.

Figure 1.5.8: Left, unusual copper tinted aragonite crystals on botryoidal calcite lightly tinted by included conichalcite. Copper tinted aragonite crystals such as these were extremely uncommon in the mines at Bisbee, 7th level, Southwest Mine, specimen - 4.5 inches.

During one of the many visits to the 7th level cave, the authors found uncommonly large amounts of conichalcite, a calcium copper arsenate hydroxide, concentrated along one wall. One small area had a few calcites which were lightly colored a greenish-yellow by the inclusion of minor amounts of conichalcite, something not seen elsewhere. This same area also hosted several of the blue-green to green copper colored formations as well (Figure 1.5.8). Also, plattnerite, a relatively common lead oxide at Bisbee, was found nearby as radiating blooms of tiny, splendent crystals several inches across. Mimetite, a lead arsenate, occurred as tiny, bright yellow crystals in one area and as a thin greenish crust in another. All of these minerals are uncommon in natural cave environments.

It was in 1927 when a detailed description of the oxidation caves at Bisbee was published. The economic geologist, Edward Wisser, published his extensive study titled *"Oxidation Subsidence at Bisbee, Arizona"* in the premier geologic journal, *Economic Geology*. Wisser's paper discussed how to recognize oxidation subsidence features and to use them as a guide to finding ore – something the miners at Bisbee had understood and employed for many years. Since by 1927 most of the oxide orebodies had been mined at Bisbee, Wisser's study was to serve as an aid in prospecting for ores in other, geologically similar mining districts.

Wisser clearly described how oxidation of the orebody caused subsidence at Bisbee. He noted cave development as an often-encountered feature associated with oxidized ores and the presence of subsidence cracks. Wisser called them "collapse caves," noting that floors of these caves were usually covered by fallen material that he called "jumbled ground," more commonly referred to as breakdown" today in cave genesis terminology.

However, while breakdown was an important aspect in the development of the caves, oxidation of the sulfide orebody and related subsidence created the conditions that allowed the collapse to occur. Without the total oxidation of the sulfide replacement deposits, the collapse would never have occurred. Thus, the authors chose to characterize these caves as "oxidation caves" a more descriptive term, as opposed to Wisser's chosen term of "collapse caves."

Wisser also noted the role of solution enlargement in cave formation and that the nature of the hosting limestone was an important factor in the cave development and related subsidence. Wisser's illustrations, Figures 13 and 14, shown here as Figures 1.5.9 and 1.5.10 illustrate the combined effects of subsidence, collapse as well as host rock bed separation and the role they play in the formation and the enlargement of an oxidation cave.

"SUBSIDENCE IN THICK BEDDED LIMESTONE. – Fig. 13 [Figure 1.5.9] *shows subsidence effects connected with an oxidized ore body in thick-bedded limestone. A collapse cave surmounts the thickest part of the ore body and beneath the cave lies the jumbled ground. Other manifestations shown are the sagging apart of the beds and a well defined marginal crack. The sagging apart of the beds is of local occurrence and immediately over one edge of the ore body.*

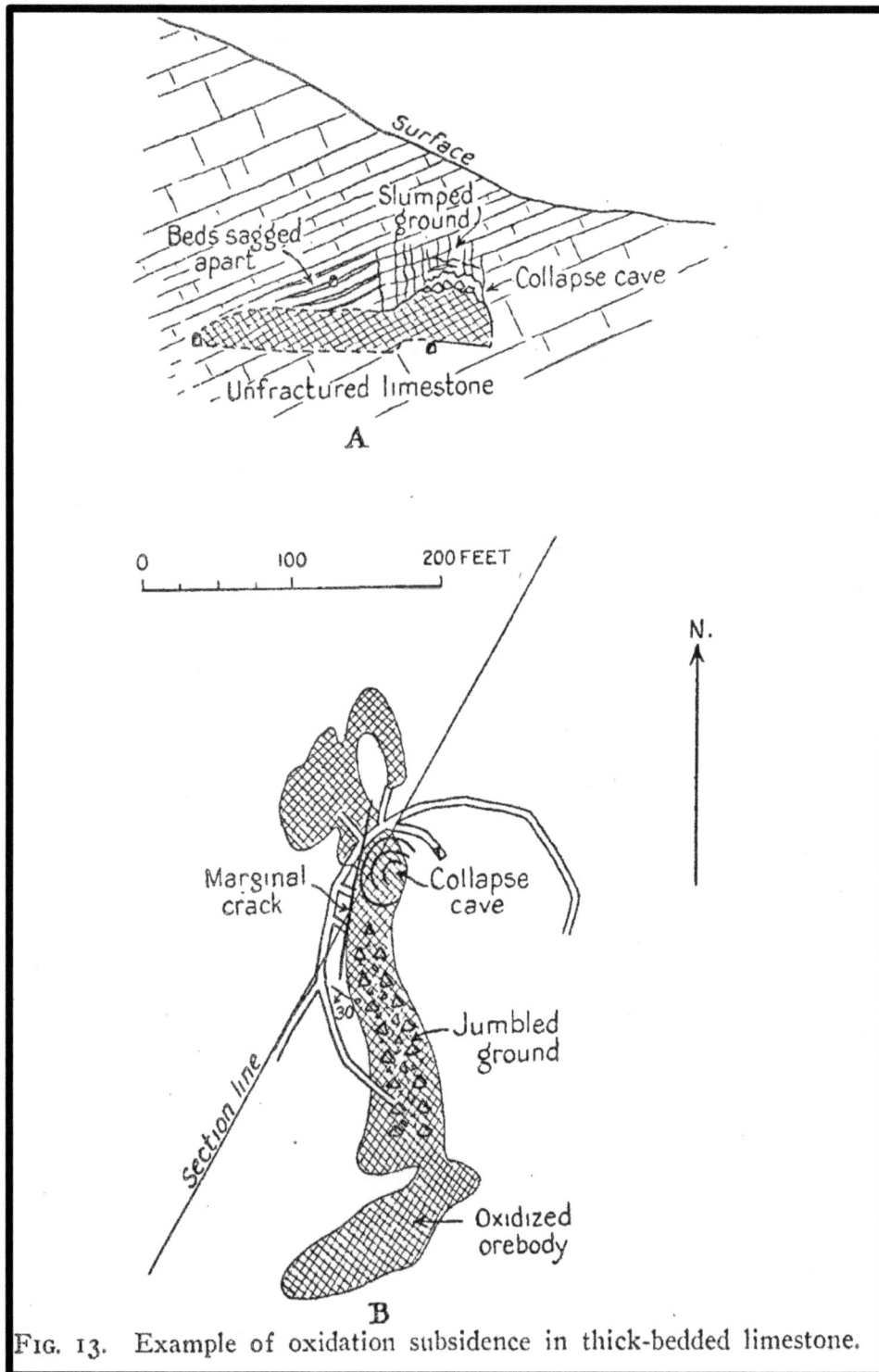

FIG. 13. Example of oxidation subsidence in thick-bedded limestone.

Figure 1.5.9: Effects of oxidation in thick-bedded limestone. Note the sag cave formation, where the beds have separated, in addition to the "collapse cave" and the linear marginal crack (Wisser, 1927).

SUBSIDENCE IN THIN-BEDDED LIMESTONE. – Fig. 14 [Figure 1.5.10] shows subsidence effects mapped above extensive ore bodies in thin-bedded limestone. While collapse caves and doming cracks exist immediately above the ore bodies, sag cracks or sag caves occur fully 90 feet above the top of the ore, illustrating the tendency of the sagging effects to climb high in thin-bedded limestone. The close association of the marginal cracks with the ore boundaries is well shown in Fig. 14" (Wisser, 1927).

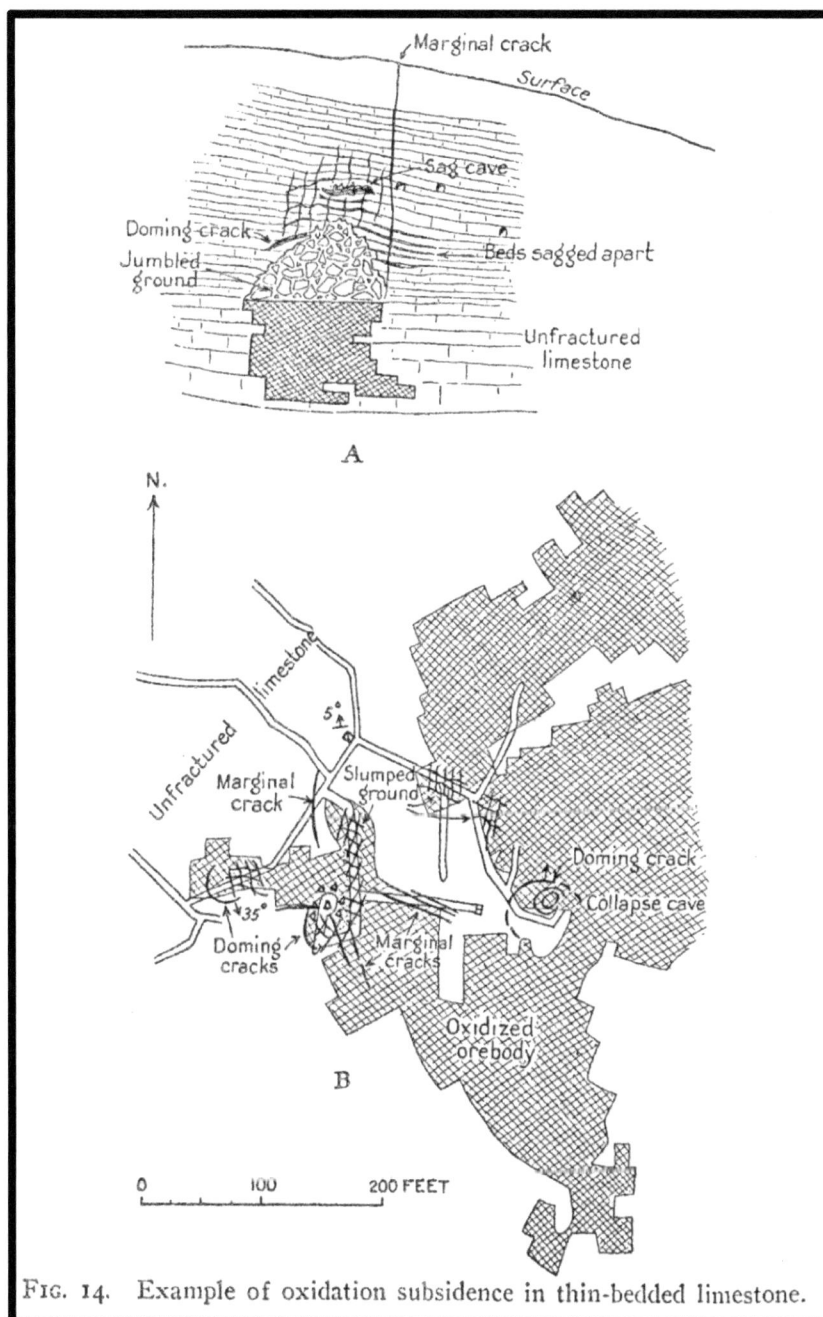

FIG. 14. Example of oxidation subsidence in thin-bedded limestone.

Figure 1.5.10: Effects of oxidation in thin-bedded limestone. Note sag cave formation and the linear marginal crack here as well, along with the arched doming cracks (Wisser, 1927).

Wisser also explained some of the other subsidence features that were often cave-like in appearance. Many of these other, typically small openings had been discovered during mining over the years, but their relationship to oxidation of the ores had been much less obvious than the caves directly overlaying the oxide ores. These features were named by Wisser as "marginal cracks," "doming cracks" and "sag caves."

Marginal cracks were defined as:

"Shear cracks, rising steeply from the periphery of the ore body to the surface…"

Vertical to near-vertical cracks were formed by the subsidence around the margins of the underlying orebody and cave. These "marginal cracks" were generally narrow features and often completely filled with banded, flowstone type calcite. It was common for such cracks to be somewhat wider near the oxide orebody because of solution enlargement or minor rotation of the subsided limestone caused by the differential settlement of the rock plug.

On rare occasion, the crack remained wide for some distance, both vertically and horizontally. The one illustrated here, as Figure 1.5.11 was open for nearly 80 feet vertically and as much as 100 feet horizontally.

On the 200 level of the Czar Mine, Richard Graeme III followed a marginal crack for more than 250 feet horizontally. This same crack was more than three feet wide and extended several tens of feet vertically. It was lined with typical flowstone type calcite. Where broken, one could see that the thickness of the flowstone calcite exceeded one foot.

To miners, open marginal cracks were a source of potential danger as they frequently contained loose rocks that were often

Figure 1.5.11: Calcite growth along a marginal crack above an oxidation cave. Minor chrysocolla and malachite development can be noted along a steeply dipping, crosscutting structure at the left, 6th level, Southwest Mine, vertical view – 6.5 feet.

hundreds of feet above and, which could fall at most anytime. Thus, when these open cracks were hit, they were always securely closed with heavy timber above, as illustrated in the Figure 1.3.3, and backfilled below as well. Doming cracks were described by Wisser as:

"...caused by the progressive breaking away of the fractured and loosened material above a weakened block by spalling and arching. ...cracks parallel to the dome itself."

Figure 1.5.12: Minor botryoidal calcite growth along a doming crack associated with an oxidation cave, "A" level, Copper Queen Mine, 1964.

Doming cracks were convex as so clearly demonstrated by Wisser and as shown by Figure 1.5.12, above. These cracks could be several feet wide and exceeded 100 feet in arced length. More than a few were beautifully decorated. Some actually graded into sag type caves as they flattened out over and become more or less parallel to the ceiling dome of the cave below.

According to Wisser, sag caves were formed where:

> *"The beds are pulled apart to form slump caves in many irregularly spaced places."*

Figure 1.5.13: **Sag cave in the Holbrook Mine, C – 1905. View dimensions are unknown, but the vertical is almost certainly less than five feet.**

While sag caves were quite abundant in Bisbee, most were small. However, a few were more than several tens of feet long and they seldom were more five feet high. Many sag caves were stunningly decorated. Colored speleothems occurred where there were oxidizing copper ores above or up dip along the limestone beds.

Since sag caves must have served as underground channels for mineral-rich water, it was common for them to be lined with thick calcite deposits. In a few instances, these deposits were crystalline, leaving the whole of the sag cave decorated with crystals including crystal overgrowths on any pre-existing stalactites and botryoidal forms. In one case, the crystalline calcite was deposited as huge rhombohedral crystals. Individual crystals exceeded 18 inches on a crystal face. The crystals were colorless to white with a frosted surface composed of many tiny, oriented rhombohedral crystals.

Boulders of the collapse-generated breakdown, accumulated on top of the oxidized ores to varying depths depending on the cave size and host rock properties. This breakdown was usually limestone, but it was typical for iron oxides and ore minerals to be mixed in as well since they commonly formed a part of the cave walls or ceilings, from which they fell. Accumulations of breakdown were quite variable, but it was not unusual for more than 20 feet of rubble to collect on the cave floor. Calcite frequently cemented the

Figure 1.5.14: Sag cave at the intersection with a marginal crack. Note the fine-grained material that has entered the cave from the surface along the marginal crack, 8th level, Southwest Mine, horizontal view - eight feet.

Figure 1.5.15: Cross sectional view of breakdown in a cave bottom composed of limestone blocks with some hematite, all cemented by calcite and resting on oxide ores of goethite with malachite. As can be seen in the photograph, the ore had been mined, leaving only a small amount next to the oxidized breakdown 100 level, Higgins Mine, horizontal view - 8.7 feet.

breakdown so completely that it could be mined under with minimal ground support. This is shown in Figure 1.5.15 below.

Often, thick layers of flowstone calcite covered and completely masked the breakdown. Several tens of feet of travertine type calcite in a cave bottom were common.

In caves that developed associated with very rich orebodies, it was common to have large amounts of copper ore minerals as a significant component of the breakdown rubble in the cave bottom. Incredible specimens of the copper and copper/zinc carbonates were found in the rubble, both as fallen material, and deposits that formed on or within the voids of the rubble, as shown in Figure 1.5.16.

When the noted mineral dealer and Tiffany and Company gemologist, Fredrick Kuntz in 1885, was describing aurichalcite (zinc/copper carbonate) at Bisbee as occurring *"in beautiful tubes lining cavities."* This was almost certainly a reference to voids between breakdown boulders in cave bottoms being lined with aurichalcite.

Figure 1.5.16: Breakdown composed mostly of hematite and goethite containing large amounts of rosasite as well as some malachite, "B" level, Uncle Sam Mine, view - 10.5 feet.

Chapter Six: The 1930s

In the late 1930s, a crosscut on the 800 level of the Cole Mine struck the middle of large oxidation cave. This was the one and only such cave ever hit in this expansive mine and was located in the western-most part of the mine, not too far from the Shattuck Mine sideline.

Richard Graeme III visited this cave in 1960, long after the ore beneath the cave had been completely mined out and partial backfilling of the cave had taken place. What remained of this cave was a space in excess of 200 feet in diameter and near circular in shape. Little of the cave was decorated with speleothems, but one massive, green-tinted stalactite approaching 20 feet in length, hung, chandelier-like, from the domed ceiling more than 50 feet above. Smaller, similarly colored stalactites were clustered nearby. The height of the cave had prevented these formations from being collected by the ever-present and opportunistic mineral collecting miners.

For the most part, the cave walls were bare limestone. In places, the limestone had been slightly altered to a light tan or cream color, streaked with black veinlets of an unidentified manganese oxide. Few of the caves visited have contained so little in the way of speleothems.

Interestingly, much of the backfilling had occurred during the years just after World War II when there was a huge excess of steel because of the war surplus of scrap. As the mining company was improving mine car types at the time, the old, end dump rectangular "A" series mine cars were loaded with waste, hauled to the cave and simply allowed to fall into the open cave. Their value as scrap metal was far less than the cost to remove the cars from the mine.

The sight of dozens of rusting mine cars in a jumbled heap was a strange one indeed. Beneath the dripping stalactites, a cream-colored crust of calcite had begun to form on several of the mine cars. This post-mining calcite faithfully reflected the shapes of the covered steel straps and rivets. An example similar to that described above is shown in Figure 1.6.1.

Figure 1.6.1: Post-mining calcite growth on a mine car, Higgins Mine, Tunnel level, 1967. More than 1/2 inch of calcite had been deposited in the just over 20 years since this mine closed. Note the calcite deposition on the left crosscut wall as well. The calcium carbonate rich water is flowing from a small diameter diamond drill prospecting hole.

The calcite on the several mine cars in this cave was already more than ¼ of an inch thick after a mere 15 years. This thickness was surprising, as both calcite and aragonite deposition rates in the oxidation caves, as speleothems, always appeared to be very slow, similar to that seen today in most any cave in the southwestern U. S.

While no actual measurements of speleothem growth in these caves were ever made by the authors, on numerous occasions the amount of deposition following the discovery of a cave by mining was obvious. This was often indicated by post-discovery calcite or aragonite deposition over the layers of dust induced by mining. In the many instances where the cave bottom had been mined then backfilled, the amount of post-discovery calcite/aragonite is quite obvious. Invariably, very little deposition had taken place, even in the earliest of the caves mined.

To be sure, mining or mining induced subsidence could disrupt, if not completely divert, groundwater flows, which would affect deposition rates. Perhaps this was the cause of slow deposition in some of the caves, but it could not have true in all instances. Thus, there is no reason to suspect that the calcite/aragonite deposition rates in the oxidation caves were any more or less than would be expected under current climatic conditions.

However, in favorable post-mining environments, the relatively rapid deposition of post-mining calcite or aragonite was commonplace. Numerous places in the Bisbee mines contained substantial post-mining calcite or aragonite, often with more than half an inch deposited within less than 20 years. The floor of 12 crosscut on the 2700 level of the Campbell Mine had more than ¾ inches of flowstone type calcite deposited for several hundred feet by swiftly flowing water in but a dozen years.

A great example of post-mining aragonite is illustrated in Figures 1.6.2 and 1.6.3 where a fragment of mine timber had from ¼ to ¾ of an inch thick coating, forming a five inch long stalactite.

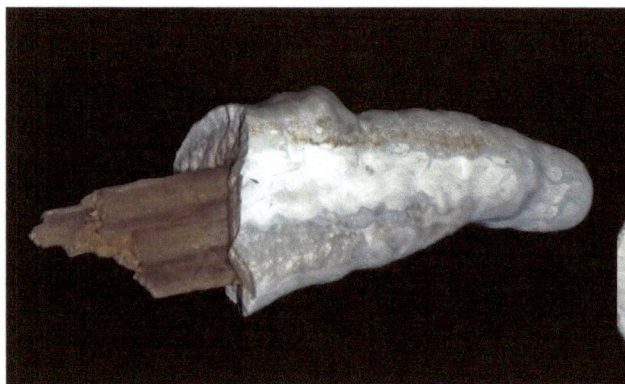

Figure 1.6.2: Post-mining aragonite stalactite formed around a fragment of mine timber, 2200 level Campbell Mine, specimen including wood - 6.5 inches.

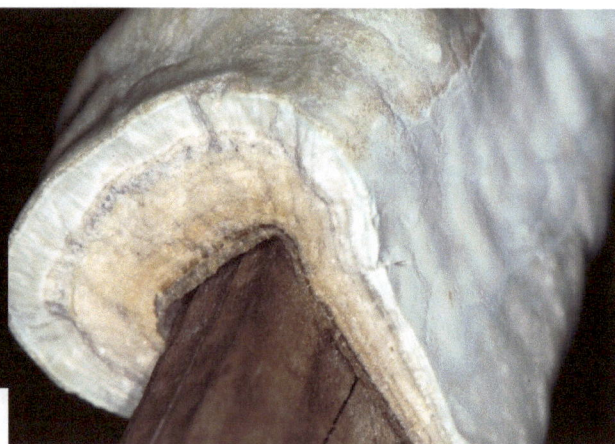

Figure 1.6.3: Close up of the specimen illustrated in Figure 1.6.2. The maximum thickness of the aragonite is 0.77 inches.

Chapter Seven: Some of the Caves Visited

The following cave descriptions cover but a very few of the many caves visited and studied by the authors over the years. Those caves selected for discussion, have been chosen because they offer examples of the more important features related to these oxidation generated openings or present an uncommon feature. Several caves, in five mines, are briefly described below.

A Small Copper Queen Mine Cave

Figure1.7.1: Right, cream-colored stalactites, white drapery with a red-brown column and flowstone in the background in a small cave some 60 feet above the Meyers Tunnel level of the Copper Queen Mine, vertical view - nine feet, 1957.

A small cave in the Copper Queen Mine was first visited in 1952. This cave was about 60 feet above the Meyer's Tunnel level of the mine and over a stoped out (mined) orebody, undoubtedly mined well before 1900.

It was accessed through the cave floor via a short raise, which cut through more than ten feet of massive travertine flowstone, a not too uncommon thickness for these cave bottoms. The cave was but 80 feet long, 30 feet across, and no more than 20 feet high and was once highly decorated with many cream-colored to brown to red-brown stalactites, two of which were more than ten feet in length.

Figure 1.7.2: Right, an uncommonly stout stalactite, measuring more than seven feet long and two feet in diameter in a small cave some 60 feet above the Meyers Tunnel level of the Copper Queen Mine, 1957.

Another very large stalactite was the thickest the authors have ever seen in any of the Bisbee caves. It was just over seven feet long, but a stout two feet in diameter (Figure 1.7.2). Large drapery type formations were abundant and made even more impressive by a combination of deep red-brown and white coloration. On one side of the cave, the formations were predominantly a deep red-brown in color and dominated by a massive 12-foot high stalagmite (Figure 1.7.3). As can be seen in this and the other photos of this cave, it had been heavily collected by miners and visitors alike.

The obviously unstable ground in this cave was something worth monitoring. Several substantial cracks exhibiting displacement crossed the ceiling. These cracks formed due to shifting of the ground during and after the removal of ore. The gradual but

Figure 1.7.3: A large, red-brown stalagmite in a small cave in the Copper Queen Mine, Meyer's Tunnel level, as it appeared in 1957. Note the crack with more than a foot of displacement in upper right, indicating post-mining movement, vertical view - 13 feet.

inevitable collapse of the cave was of great interest us. This was a rare chance for us to monitor the collapse of such an opening. By the early 1950s, several large boulders had already fallen from the ceiling and the movement of the ground, though slow, was perceptible over the course of several years. During the late 1980s, an effort was made to visit the cave, but it had completely collapsed.

A Copper Queen Mine Cave with Rich Ore Still in Place

In mid-1954, Richard Graeme III was on the 400 level of the Copper Queen Mine, near the bottom of the Copper Queen Inclined Shaft, when he noticed an oval shaped hole near the top of the crosscut. Holes of this type were common throughout most of the mine workings, particularly near oxide orebodies. Few held anything of note and this one, lined with cave type calcite, showed little promise. While the opening was much too small for an adult to enter, a 12-year-old boy could and he did. He crawled for perhaps 40 feet along an 18 to 30 inch high calcite-lined crack in goethite until it opened up into a small cave.

While the cave was well decorated, it held nothing exceptional, so he left. The exceptional was to be found on the way out where, in spots, the thin veneer of calcite had been scraped off the rock on the floor during the crawl on the way in. Just under the thin cave type calcite was malachite - an incredible mass of chatoyant malachite. It extended for at least six feet, before it disappeared under the fallen rock of the cave floor. In places, it was up to four inches thick and sitting on soft goethite.

Little effort was required to remove the malachite as the malachite easily separated into manageable size pieces. This natural separation into smaller pieces is a feature common to malachite, something caused by the gel manner in which it forms. The surface tension of the gelatinous masses during formation often prevents the individual, adjacent masses from growing together, leaving natural separations. Several hundred pounds of high quality malachite were recovered from along the crawl hole.

Figure 1.7.4: Cave-type calcite partially coating malachite with goethite. Copper Queen Mine, view – five inches. This specimen is very similar to those described, but is not part of that collected by the author in 1954.

What else may have existed in the adjacent areas is difficult to say, as only a limited amount of exploration work was completed before the single access was blocked by Phelps Dodge as a necessary safety precaution. A few years later, a fire in the Copper Queen incline shaft completely sealed the lower portions of the shaft and any access to this area.

This cave occurrence of such malachite was of the type that was common in the richest of orebodies mined during the earliest of days at Bisbee. This represented one such orebody, which had not been found, though there is no reason to believe it was of any great size.

Caves Associated with the New Southwest Orebody, 6th and 7th levels, Southwest Mine

The New Southwest Orebody was discovered in 1919 and yielded well in excess of one million tons of ore during mining over the next nine years. While much of the open area was backfilled, waste rock was scarce in the upper levels of the mine and the backfilling was never completed. A recipe for certain disaster as such large open areas were always unstable. By 1932 there was little activity in this expansive mine. Phelps Dodge had leased several limited areas of the Southwest and other mines to small time miners, locally called "lessors" (Mills, 1958). Only a few lessors were now working in scattered locations in the Southwest Mine. Then, during early 1933, a sudden, catastrophic collapse of several contiguous stopes occurred. Fortunately, it occurred late at night and no one was in the mine at the time. A number of smaller collapses followed over the next several years. This also caused the overlying caves to collapse. Soon, the associated cracks and subsidence reached the surface as the whole of the area settled following the cave in. Safe access to the area was lost from the 4th level to well above the 7th level.

Figure 1.7.5: A small section of the Ballpark Stope with cave-type calcite coated boulders from a cave bottom, which dropped into the collapsed stope, 6th level, Southwest Mine, horizontal view in the middle ground – greater than 60 feet.

The resulting void was a huge seemingly impenetrable, vast darkness, filled with house-sized boulders. The spaces between the boulders were canyon-like; both deep and wide. The most powerful of mine lights failed to pierce the enveloping blackness.

Indeed, any light seemed to be completely absorbed by the surrounding darkness. The miners of the day aptly name this immense opening the "Ballpark." Few if any ever entered for fear of further collapse (see Figures 1.7.5 and 2.1.5).

In the early 1970s, one of the authors was conducting a systematic, district wide sampling of rock alteration types for the mine owners when he came back to the edge of the Ballpark for the first time in most of 20 years. Eager to explore, he started into the foreboding vastness, but soon found himself alone, as not one of the accompanying geologist or miners followed. The other members of the party would not venture in, so intimidating was the size and the appearance of recent collapse. No more than a hundred feet were explored before the danger of proceeding alone became obvious. There was an overwhelming sense of quickly becoming disoriented and soon lost, thus exploration of the Ballpark was postponed until it could be done safely, systematically and with others.

A few years later, the authors returned to explore the Ballpark, together, and spent several hundred hours over the next years exploring this vast underground space. The deep cracks between the huge boulders were explored and every space under the large rocks entered. Sometimes crawling for several hundred feet, just in these interconnected spaces, small caves, totally contained in a huge boulder would be found. Numerous small and several reasonably large remnants of caves were found in and above the collapsed area of the Ballpark. These included caves formed along doming and marginal subsidence cracks, now made wider by the massive collapse; sag caves exposed due to the collapse and a couple of typical oxidation caves, which had formed over the New Southwest Orebody. In all cases, the caves were almost totally destroyed by the extensive caving of ground. Cracks extended more than a thousand feet horizontally and several hundred feet vertically and actually reach the surface. A

Figure 1.7.6: Extraordinarily large botryoidal forms of calcite. The largest bulb shaped form extends 3.8 inches above the crystalline calcite base. These have an uncommonly high luster and, as can be seen, a light pinkish tint. Because they were not in place when found, the orientation during growth could not be determined, 6th level, Southwest Mine.

number of interesting cave formations were found in the domed ceiling of the collapsed stope and in the chaotic mass of gigantic boulders which now formed the irregular floor of the immense opening.

Also, fallen speleothems were found among the collapse floor material in literally dozens of spots. Some were largely intact, but most had been shattered by the fall. A number of representative examples of cave formations of unusual size or form were recovered from the fallen material for study and preservation.

Figure 1.7.6 is an example of a botryoidal calcite, with unusual large bulb-shaped forms, from the collapsed rock on the floor of the Ballpark. Interestingly, the calcite crystals associated with these are near razor sharp. Specimen recovery was a painful exercise because of the many cuts received by simply picking up the sharp rocks.

Other examples of unusual calcite forms and one aragonite in classic flos flori form are illustrated in Figure 1.7.7. Most of the examples illustrated, manifest an unusual form from a late stage, subaqueous overgrowth of crystalline calcite. This is because the several caves in which they developed were completely or partially filled with calcium carbonate rich solutions at some point.

Specimens one and five formed on the tip of stalactites as the solution level reached that point. Calcite crystallization took place at and below the solution level in a radiating fashion, using the calcite of the stalactite tip as a nucleus. Specimens numbers two and three represent different crystal growths over pre-existing stalactites. In specimen two, the stalactites became large, single crystal speleothems. While with specimen three, rhombohedral calcite crystals developed more or less perpendicular to the stalactite core. In the case of number six, normal botryoidal calcite was completely over grown by small calcite rhombohedral crystals, here too; they are perpendicular to the underlying calcite surfaces.

Figure 1.7.7: A group of speleothems recovered from the collapsed caves of the New Southwest Orebody.
1) Horizontal calcite growth on a stalactite tip reflecting a water line at this point for an extended period. Specimen - 3.3 inches wide, 7th level.
2) Single crystal stalactites on a large rhombohedral crystal base. Specimen – 10.8 inches long, 7th level.
3) Calcite stalactite completely overgrown and terminated by rhombohedral calcite crystals. This reflects submersion after formation of the stalactite. Specimen – 8.4 inches long, 7th level.
4) Flos ferri variety of aragonite. The individual erratic forms are hollow. Specimen – 8.5 inches, 8th level.
5) An incredible circular cluster of calcite crystal which formed on the tip of a stalactite at the water line. Specimen - 3.6 inches, 6th level.
6) A cluster of botryoidal calcite forms coated by an overgrowth of calcite crystals caused by submersion after formation of the botryoidal calcite. Specimen – 5.3 inches, 6th level.

In a number of locations, small areas of speleothems were still in place, though invariably somewhat damaged by the collapse of the surrounding rock. Examples of two are shown in Figures 1.7.8 and 1.7.9.

Helictic forms, as shown in Figure 1.7.8 were reasonably common in place. None of these delicate forms survived the fall to the Ballpark floor, as did the more durable pieces.

Figure 1.7.8: Right, several nests of helictic calcite forms along a marginal crack. The portion of the cave to the left had dropped more than 40 feet, 7th level, Southwest Mine, vertical view - ten feet.

Figure 1.7.9:   Below, multiple solution levels as shown by both iron staining and variable type calcite deposition along this doming crack cave. The lower-most calcite formed as sharp rhombohedral crystals of more an inch in size with a distinct ledge reflecting a long time water line. Above this is calcite crystal coated, botryoidal calcite, followed by red-brown iron staining of varying intensity, 7th level, Southwest Mine, horizontal view - 8.5 feet.

The Cave in Goethite on the 4th level of the Southwest Mine

The cave with the high malachite content in the stalactites discussed in chapter four, was of additional interest because it developed almost totally in goethite. While, as previously noted, it was common to have at least some of the oxidation cave occur in iron oxides such as goethite. Near total development as shown in Figures 1.7.10 and 1.7.11 was uncommon and not seen on this scale elsewhere.

Speleothems were scarce in this cave and generally occurred in the upper most portions of the cave. Only a few formations were copper tinted by included malachite (Figure 1.7.12) while others were deeply colored by iron (figure 1.7.11).

Figure 1.7.11: One of the authors, descending from the upper part of the cave to the 4th level. Note the top of the 4th level drift at bottom center. The calcite at the top is largely white, while that in the center is colored a deep brown. Some limestone is visible as rounded knobs in the left center foreground.

Figure 1.7.10: Calcite and aragonite on goethite in the upper part of the cave. Note the multiple, parallel, horizontal lines in the calcite at the right, evidence of varying solution levels – bottom left back and the unusual multiple, and parallel calcite crystal growth on the right foreground. Horizontal view – six feet.

Figure 1.7.12: Right, calcite speleothems, a few tinted light green on massive goethite, also in the upper portions of the cave.

99

A Sag Cave on the 4th level, Southwest Mine

A vertical exploration working designated as 3-14 raise was driven between the 4th and 5th levels of the Southwest Mine and clipped the edge of a sag cave about 20 feet above the 4th level. The cave was never more than five feet in height and six feet wide at most, but extended for more than 30 feet horizontally. Calcite growth gave the impression of several chambers. The sag cave was particularly interesting because over its pre-mining existence, it had experienced several episodes of being flooded with calcium carbonate rich solutions.

In the lowest section of the cave, the whole of the ceiling was composed of enormous rhombohedral calcite crystals, many of which were well over a foot long. Larger crystals, some measuring more than 18 inches across a crystal face, were scattered among an impressive array of smaller, translucent crystals. An example of a large composite crystal is shown in Figure 1.7.13.

Several levels of crystalline shelfstone development (tiered), suggest that over time the solution levels were at different levels. Later flowstone and stalactites told of a more recent time, with typical speleothem development found in

Figure 1.7.13: Composite, calcite rhombohedral crystal, 4th level, Southwest Mine, specimen - nine inches tall.

an open air environment. Locally, flowstone partially covered both shelfstone and the large calcite crystals. In some cases, this white flowstone hosted gnarled clusters of copper tinted calcite helictites.

An occasional, deeply copper tinted stalactite hung in marked contrast to the more common white formations covering the ceiling and walls (see Figure 1.7.14). The huge calcite crystals and the intensely green colored stalactites made this a most impressive cave, unlike any other seen by the authors.

100

Figure 1.7.14: Calcite stalactite intensely colored by copper, 4th level, Southwest Mine, vertical view - 21 inches. Note the coarsely crystalline nature of the calcite, as indicated by the pattern on the side of this translucent stalactite.

A Cave on the 6th level of the Southwest Mine with Unusual Speleothems

A small, goethite/hematite hosted cave had been bisected by an exploration crosscut (Figure 1.7.16). While much of the cave had been backfilled with waste rock from the crosscut, one area of the cave was still open. For the most part, everything in and about this cave had been destroyed during mining and the remaining, five foot high and ten foot by 22 foot opening had been used for the disposal of trash including dynamite and blasting caps, an uncommon and strictly forbidden practices at the time. However, this opening and a smaller adjacent room contained aragonite in a form never seen before anywhere.

Figure 1.7.15: Two of the authors in the crosscut that intercepted this cave. The whole of entire ceiling is the top of the oxidation cave, as shown by the scattered calcite speleothems. The left portion of the cave was completely backfilled while the right portion was largely open. Note the goethite and hematite wall on the right.

Helical growths some more than ten inches in length hung in small erratic clusters from the ceiling. Several single growths protruded horizontally from the calcite-covered walls. A few of the helical growths were exceptionally sharp in form, reminding one of machine screws as can be seen in Figure 1.7.16. Interestingly, both right and left-handed twist were present in most of the clusters with right-handed twist far more common.

Figure 1.7.16: Specimens recovered from a 6th level, Southwest Mine cave.

Left, a cluster of helical aragonite growths with minor brown calcite overgrowth on the longer forms. Specimen - 18 inches.

Right, a sharp helical aragonite growth with calcite on one side and specks of rosasite. Specimen - four inches.

Pencil-lead thin, goethite stalactites, some exceeding eight inches in length, had formed as scattered clusters in one area of the cave. Localized calcite overgrowth on the goethite stalactites resulted in attractive groups of very thin calcite stalactites.

At some point in its history, the cave had been at least partially filled with low pH, very high iron solutions. This would not be too surprising given the cave formed in goethite, a mineral that forms under these conditions.

Evaporation of the solutions caused the development of thin, floating rafts of hematite and goethite on the solution surface, just as so commonly occurs with calcium carbonate rich water. As they became larger and heavier, they sank, similar to calcite raft formations. The authors have seen this development of hematite/goethite rafts in other locations at Bisbee on low pH mine waters.

The resultant accumulated mass of mixed hematite/goethite rafts was several feet thick (see Figure 1.7.17). As these sunken rafts continued to grow below the solution level, they acquired a strong iridescence (Figure 1.7.18). This iridescent, hematite/goethite mixture is known as "turgite" in the science of mineralogy. Because turgite is a mixture of two defined minerals, it is not considered a valid mineral species, in spite of the wide use of the name.

Figure 1.7.17: An accumulation of hematite and goethite rafts, some 18 inches thick in the cave on the 6th level, Southwest Mine.

Figure 1.7.18: Iridescent hematite/goethite (turgite) raft litter cemented together with calcite, 6th level, Southwest Mine, specimen - 3.8 inches.

A Cave Filled with Gypsum Speleothems on the 6th level Southwest Mine

A small oxidation cave on the 6th level of the Southwest Mine was most unusual in that almost all of the speleothem were gypsum. Few calcium carbonate speleothems were present at all in the cave.

The opening was no more than 40 feet long by 15 feet wide and never any more than eight feet in height. It had developed surprisingly close to the surface and over a modest sized copper/ zinc orebody in manganese rich Martin Limestone. The almost-white gypsum was a stark contrast against the near black walls of the altered, high manganese limestone and the similarly colored breakdown.

A crust of gypsum to one half inch thick coated most surfaces in the upper part of the cave. The breakdown on the cave bottom was loosely cemented with thicker growths of spongy gypsum. In the small voids between these boulders were the occasional ram's horn type of growths, usually as single growths to a maximum of four inches in total length (Figure 1.7.19 and 2.7.12). A few clusters of such growths were noted however.

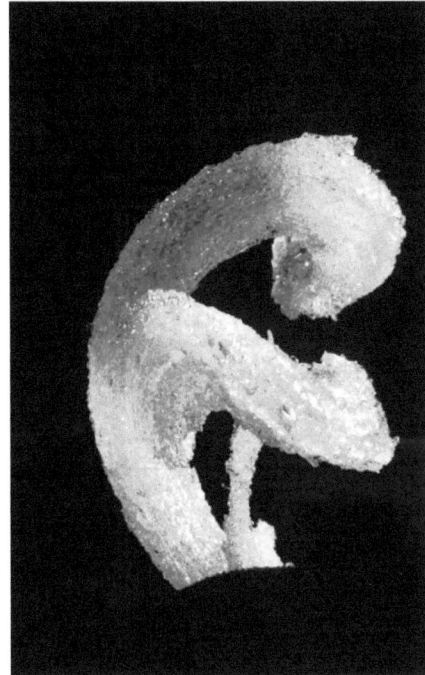

Figure 1.7.19: A 3.5 inch high gypsum ram's horn, 6th level, Southwest Mine.

Spongy, lightweight blocks of intergrown selenite crystals (a crystalline variety of gypsum) were everywhere. Some of these strange, unconnected blocks of gypsum exceeded 18 inches in the greatest dimension. While the blocks were largely white, black romanèchite often speckled the top surface. Malachite occurred as occasional, tiny acicular crystals scattered randomly in and on the spongy gypsum. Less common were equally small, rounded blebs of rosasite (Figure 1.7.20).

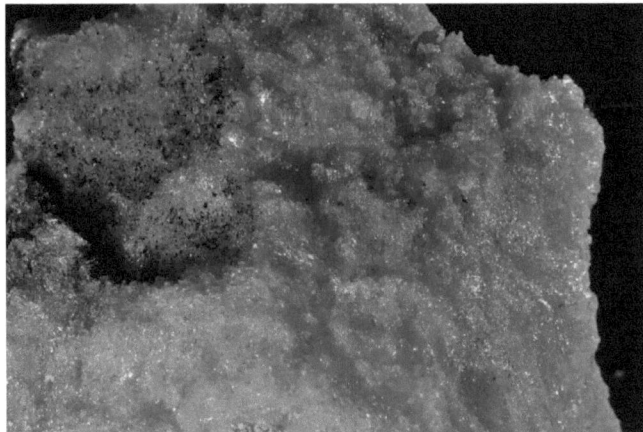

Figure 1.7.20: A spongy block of tiny selenite crystals with black romanèchite, minor malachite and a few isolated specks of rosasite, 6th level, Southwest Mine, horizontal view - six inches.

The lowest part of the cave floor was covered with closely spaced, colorless half-inch high selenite crystal, all oriented perfectly vertical. The seven-foot wide area had obviously been a pond when these crystals grew. All of the surrounding gypsum was a brilliant white, alabaster-like material.

104

A Cave in the Uncle Sam Mine with Copper/Zinc Carbonate Minerals

South of the Czar Mine and on the other side of Queen Hill is the Uncle Sam Mine, situated between the Cuprite Mine and the exceptionally rich Shattuck Mine. The Uncle Sam was never a great mine like most of the others owned by the Copper Queen. But for a few years, beginning in 1910, the Uncle Sam Mine was an important producer (Mills, 1959), exploiting a few rich orebodies, including some oxide ores associated caves such as the one noted in chapter four.

While on the "A" level of this mine, the authors found a solution enlarged, subsidence marginal crack which had been intercepted by an exploration crosscut. One wall of this crack had a thin layer of goethite with some rosasite mineralization (Figure 1.7.21). Much of the rosasite was actually an incomplete replacement of malachite pseudomorphs after azurite as shown in Figure 1.7.22.

Figure 1.7.21: Rosasite replacement of malachite on goethite along a marginal crack. "A" level, Uncle Sam Mine, view - 3.5 feet.

Figure 1.7.22: Rosasite replacing malachite pseudomorphs after azurite with minor, tiny colorless hemimorphite crystals. "A" level, Uncle Sam Mine, specimen - three inches.

Figure 1.7.23: Aurichalcite on cave wall, "B" level, Uncle Sam Mine, view - 3.5 feet.

Less than 30 feet along the narrow crack, was a calcite-lined opening that led some 20 feet downward toward the "B" level and into a small cave. It was obvious that no one had ever entered the cave before as small, but sharp calcite crystals on the floor were broken with every inch of difficult, forward progress.

The cave was almost totally developed in mixed manganese and iron oxides. In the upper part, a thin crust of calcite coated the walls and floors, but this soon gave way to barren, pulverulent manganese oxides. Black dust quickly filled the air with any movement through the cave, but the dust did not hide the lovely patches of delicate aurichalcite (zinc copper carbonate) crystals covering the walls for several feet in places (Figure 1.7.23). Tiny, colorless hemimorphite (zinc silicate) crystals were noted in small amounts in a few spots on the cave walls as well. The rubble on the floor was composed of vuggy, goethite/manganese oxide boulders, rich with malachite, aurichalcite and botryoidal rosasite (copper zinc carbonate), though somewhat obscured by a thin layer of black/brown dust. An example of these boulders is shown in Figure 1.5.16.

A Cave with Very Different Formations in the Shattuck Mine

The Shattuck Mine hit a number of interesting caves as documented several times in chapter four. However, many, many others went unrecorded as was so typical for the era. During the early 1970s, Richard Graeme III was part of a team responsible for reopening the Shattuck Mine. The mine had been closed some 30 years before. All of the surface structures and the shaft timber were burned in fire accidently set by children in 1952 (Mills, 1956). During the course of this activity, he came to know the Shattuck Mine well, including the location of a number of scattered and rather ordinary oxidation caves. In the late 1970s, the authors began exploring the mine together, as a group and, over time, found several more oxidation caves, including the one described herein.

Figure 1.7.24: Botryoidal hematite with a later, partial dusting of specular hematite and minor calcite, 600 level, Shattuck Mine, view - 17 centimeters.

This was one of the more unusual Shattuck caves and accessed through the 600 level of the mine. The cave had formed adjacent to and partially within a mass of many thousands of tons of primary hematite, something not seen in any other cave.

While both primary and secondary hematite is a common mineral at Bisbee, this was a most unusual occurrence of primary hematite because of its large size and relative purity of the hematite (Figure 1.7.24).

Figure 1.7.25: Calcite crystal cluster on a cave wall, 700 level, Shattuck Mine, horizontal view - 5.7 inches.

The cave was a near vertical opening, which formed along a marginal subsidence crack, which had developed along the interface between the hematite mass and the limestone host rock. The opening was well over 100 feet in height and nearly as long, but no more than ten feet at its widest point. The decoration of this small cave was unique.

Nearly all of the walls were covered with odd, branching clusters of calcite crystals, colored a deep red/brown by the hematite. These delicate clusters were up to six inches in composite length and all were oriented nearly horizontal, or more or less perpendicular to the near perpendicular cave wall, as shown in Figure 1.7.25. A closer look at the crystals in Figure 1.7.25 will reveal a dull calcite deposit on the upward facing crystal surfaces. The crystals were deceptively sharp as Richard Graeme III learned as he grabbed the wall for support when a timber at the edge of the backfill broke under his weight.

This Shattuck cave extended through to the 700 level, where mining had removed the ores beneath. While much of the cave had been backfilled, a small portion of the preserved cave bottom was covered with thick, billowy and very white calcite. This white calcite was a striking contrast with the clusters of red-brown crystals grew on the cave walls.

A Classic Oxidation Cave on the 100 level of the Holbrook Mine

During the late 1960s, work in the Holbrook Extension to the Lavender Pit Mine, began to expose what was once a part of the underground workings of the incredibly rich Holbrook Mine. As the pit deepened, numerous old underground openings, now exposed in the pit walls became accessible. Most of the Holbrook Mine had been inaccessible for decades because of bad ground conditions. The newly exposed underground workings in the pit wall were little better, usually only accessible for a few tens of feet before complete collapse blocked any further access. There was one exception.

For a period of time, the authors could explore one such opening that led to a series of mid-sized stopes in the Holbrook Mine on the 100 level and underneath Queen Hill. Above these stopes was an approximately 175 feet high oxidation cave, which had not been mined or backfilled. The cave was not damaged or destroyed because the copper content in the rock immediately below the cave was not economical at the time. Also, as the cave was very high in the mine, there was no access above it for the introduction of fill, which surly would have occurred had it been possible.

This Holbrook cave was not large. It might have been 160 feet by 90 feet, but it was unusually tall. The cave was a good example of the classic oxidation caves found very early in the history of mining at Bisbee. While it probably would not have generated much interest at the time it was discovered, this cave contained impressive features. Large areas of the walls were covered by malachite more than an inch thick, particularly in the upper portions. In the lower parts, this malachite

Figure 1.7.26: Calcite and aragonite helictite cluster of some two feet in overall length, on the cave wall. The calcite on the right wall has covered malachite and chrysocolla, while the aragonite to the left is tinted copper, 100 level, Holbrook Mine.

coating was masked, in part, by calcite flowstone and, in part, altered to chrysocolla. Two foot long, malachite stalactites hung in an alcove-like pocket, high above the cave floor, well out of the reach of the miners of the era who would certainly have removed them, had they the opportunity. Perhaps they never saw these malachite stalactites, as they worked by candlelight, which would not have illuminated something so far above.

Chrysocolla was abundant as large patches in the walls, and in places, the chrysocolla coated what appeared to be massive cuprite, which was common in the oxide ores of this area. The authors were not able to confirm the cuprite substrate on which the chrysocolla was deposited, but experience elsewhere in this mine supports this supposition.

Figure 1.7.27: Oxidation cave with abundant malachite and chrysocolla on the walls and calcite speleothems deeply colored by copper and iron. Minor amounts of blue-green, copper tinted aragonite are visible in the upper portions as well.
Note one of the authors sitting on top of a stalagmite in the bottom center of the photo. Though not apparent in the photograph, he is more than 150 feet above the cave bottom. Also, there is a pocket containing malachite stalactites in the upper-center-left of the photograph, 100 level, Holbrook Mine.

Calcite and aragonite speleothems vividly displayed the coloration imparted by both copper and iron minerals. Stalactites, some exceeding ten feet in length, were variably tinted by copper while numerous smaller ones were deeply colored green to blue green (Figures 1.7.27 and 1.7.28). Some were even semi-translucent, further enhancing their loveliness.

Large groupings of fanciful helictites were scattered about the cave. In places, these too were colored a delicate blue-green by copper. Often, a later overgrowth of coarsely crystalline calcite containing flecks of malachite covered these erratic helictites, indicating that a period of submersion by mineralized water had occurred to a high level in the cave. This can be seen on the lower portions of the helictites in Figure 1.7.26 where these crystalline calcite overgrowths, covering the bottom third, can be seen.

Near the bottom of this Holbrook cave, light brown calcite crystals to six inches long, stood spear-like, in random orientation on darker brown calcite (Figure

Figure 1.7.28: Copper tinted calcite and aragonite speleothems and malachite on cave wall, 100 level, Holbrook Mine, view - 13 feet.

Figure 1.7.29: Iron tinted calcite crystals of up to six inches in length, 100 Level, Holbrook Mine.

1.7.29). These too, told of a period when calcium carbonate rich solutions stood in the cave.

On close examination, several interesting mineral species were noted in relatively minor amounts, on the cave walls. Small, scattered spheres of rosasite were found on goethite in several places. Additionally, tiny clusters of radiating, colorless crystals of hemimorphite were commonly associated with the rosasite (Figure 1.7.30).

Near the rosasite, plattnerite, a black lead oxide, was overgrown by calcite producing black calcite crystals (Figure 1.7.31). Elsewhere in the cave, plattnerite was found as exceptionally large crystals, for the species, on calcite and goethite (Figure 1.7.32).

Figure 1.7.30: Rosasite spheres to 0.30 inches on goethite. Minor, colorless hemimorphite crystals to 0.12 inches are associated with the rosasite, 100 level, Holbrook Mine.

Figure 1.7.31: Black plattnerite included in calcite on the cave wall (upper and center left), with minor chrysocolla and malachite, 100 level, Holbrook Mine, horizontal view - 20 inches.

Figure 1.7.32: Silvery-black plattnerite crystals to 0.05 inches on calcite and goethite in the background, 100 level, Holbrook Mine, view - two inches.

PART 2: SULFIDE OXIDATION AND SPELEOGENESIS AT BISBEE, ARIZONA - A REVIEW OF OXIDATION CAVE DEVELOPMENT.

The following discussion is oriented toward the more technical aspects of oxidation cave development at Bisbee and is presented to buttress our position that these openings totally owe their existence to the complete oxidation of the sulfide replacement deposits. Because this is a more technical section of the book, all measurements are presented in metric format, as is the tradition. English units of measurement were given in the historical review to be consistent with the quoted literature.

Introduction

It is a widely recognized and long accepted fact that sulfide mineral deposits have been emplaced in pre-existing caves. Examples abound throughout the world of paleokarst environments hosting sulfide ores (Morehouse, 1968, Palmer, 1991, Quinlan, 1972, Walker, 1928). Examples in the U. S. are famous mineral deposits at Tintic, Utah (Tower & Smith, 1899, Walker, 1928), Aspen, Colorado (Maslyn, 1976) and Leadville, Colorado (Emmons, et al., 1927, Loughton, 1926). The deposition of sulfide minerals in hypogenic caves (hydrothermally developed) is equally well documented and accepted with the extensive lead/zinc deposits of the Picher Field in Oklahoma and Kansas (McKnight & Fischer, 1970) as but one example of many. Further, it is also generally accepted that, in more than a few instances, supergene activity has caused the development of open caves through the oxidation of sulfide ores hosted in these paleokarst and hypogenic environments (Osborne, 1996, Tower & Smith, 1899, Walker, 1928). These resultant caves often bear some strong similarities to the oxidation caves at Bisbee, including oxide mineral stalactites (Tower & Smith, 1899).

This study has been undertaken in the context of the fine works of numerous authors on this subject and/or related ore deposits. Among the principal works reviewed are: Bosák (1989), Curtis (1884); Dźulyński and Sass-Gustkiewiez (1989); Emmons (1886); Jones, et al. (1967); Lindgren and Loughlin (1919); McKnight and Fischer (1970); Morehouse (1968); Nolan (1962); Osborne (1996); Palmer (1991); Quinlan (1972) and Walker (1928). Numerous other, but less relevant, papers were reviewed as well. Additionally, extensive underground fieldwork was undertaken in the Tintic, Utah; Eureka, Nevada; Leadville, Colorado; Hanover, New Mexico and other carbonate hosted mining areas, all of which have abundant caves associated with the oxide orebodies.

Understanding the depositional controls for the hypogene ore emplacement is absolutely essential to correctly interpreting the possible presence of pre-mineral open spaces of any origin. More than one paleokarst depositional environment was originally described as a carbonate replacement deposit (Emmons, 1886, Lindgren & Loughlin, 1919). A close and detailed study of the ore deposits at Bisbee finds no reason to suggest that either a paleokarst or a hypogenic cave environment were present or played any role in the deposition of the limestone hosted sulfide deposits at Bisbee. The many scattered and often isolated sulfide masses in the several Paleozoic carbonate horizons at Bisbee are replacement in origin.

As for the oxidation caves, it is the authors considered opinion that Bisbee's oxidation caves formed near the surface where the complete oxidation of large sulfide mineral deposit occurred in favorable host rocks such as relatively pure limestone and dolomites. Other, more typical cave forming mechanisms, such as host rock removal by mildly acidic waters of meteoric origin, played only a relatively minor role in the development of this type of opening.

As sulfide minerals are exposed to near-surface conditions, a multi-stage process takes place that breaks down the sulfur bearing minerals, liberating acidic, metallic sulfate solutions that dissolve portions of the host rock and transport some of the components of the original sulfides. When this process is able to proceed to completion, most, if not all of the sulfur has been removed and a significant portion of one or more of the original metallic components of the sulfides usually iron, have been somewhat transported away from the original depositional site. The end result of complete oxidation is a substantially reduced volume of material and an enlarged, open area partially filled by the relatively small amount of remaining material. When the host rock is sufficiently competent to develop a dome, a cave is formed. Less competent units that cannot support an opening, tend to collapse and fill the resultant space with typically angular material as illustrated from Wisser (1927) in Figure 1.5.10.

As the genesis of these caves is totally dependent on the complete oxidation of sulfide mineral deposits, the term "oxidation cave" is used to describe theses openings. Oxidation caves as found at Bisbee are typically single-chamber caves and cave-like openings. The caves are immediately above and often partially within the oxidation products. The separating of the host rock beds developed openings as they sagged due to the removal of support during oxidation of the sulfide orebodies. These too helped create the principal void and/or developed secondary voids or "sag caves."

Chapter One: History

The wonderfully rich copper mines at Bisbee, Arizona were worked from early 1880 until mid-1975. Mining operations ceased only because it became too costly to mine, not because the mineral deposits were exhausted. Mining started at the surface in thoroughly oxidized ores in the western part of what became the Warren Mining District at an elevation of about 1,650 meters above sea level. It ended in the deep, unaltered sulfides in the eastern part of the district, 1,220 meters lower, at an elevation of 430 meters, and most of four kilometers to the east of the original discovery point. Over the 95 years of mining in Bisbee, more than 3,500 kilometers of underground workings were dug by miners into the limestone and dolomitic rock units. These rocks, which held the many hundreds of scattered replacement deposits. In all, the mines produced nearly 4 million tons of copper, hundreds of thousands of tons of both zinc and lead, 100 million ounces of silver and more than one million ounces of gold (Graeme, 1981). By any standards, this was a world-class mining district, one that fed much needed, basic materials to the industry of a growing nation at a critical time.

The association of caves with many of the orebodies mined early at Bisbee was well known and documented at length in the first section of this book. The caves occurred directly over bodies of thoroughly oxidized copper ore and less commonly, oxidized lead ore. All were associated with abundant iron oxides – often tens of thousands or, on rare occasion, hundreds of thousands of tons of soft iron oxides surrounded by limestone (Figure2.1.1). Some caves were also found above huge masses of just iron oxides with little copper ore, which is a reflection of the high iron and low copper mineralogy of the original sulfide replacement deposit. As has been suggested, more than caves were formed by the oxidation process.

Figure 2.1.1: A face of soft oxide (top) on hard limestone (bottom) in the Czar Mine about 1895. The oxide was so soft it could be removed by just using a pick; no explosives were needed, except to blast the hard limestone. A consequence of this abundant clay, was the need for substantial, heavy timber to support the opening, Douglas (1900).

The same processes that oxidized the sulfide ores and formed the oxidation caves also created huge masses of soft, claylike materials that were mixed with the iron oxides surrounded the oxide ores. These clays-like masses responded plastically under the pressure of the overlying rock, putting tremendous force on any opening. Even thick timber could not support the pressure of the moving clays (Figure 2.1.2).

The difficulty of mining under these conditions was briefly noted by the *Bisbee Daily Review* in 1904 with:

"The ore bodies are so soft that the problem of timbering is a serious one, the mine requiring about thirty feet of timber, board measure, for every ton of ore taken out, and it is frequently necessary to bulkhead the openings to them to keep them intact. [See Figure 2.1.3] *The whole mountain under which the mines are opened, seem to be creeping, but despite the threatening circumstances under which the ore is taken, the exceeding care exercised in timbering, renders the mine safe, and great accidents due to the caving in of the ground have never occurred."*

With this continual "creeping" referred to above, the subsidence generated cracks in the hard, overlying limestone spread open (Figure 2.1.4), and penetrated deep into the ground, often for several hundred meters as documented by Wisser (1927). Indeed, the whole of the area was in a more or less state of constant movement because of ore removal and the plastic behavior of the claylike oxidation products surrounding the ores.

Figure 2.1.2: Typical collapse of mine workings caused by the crushing pressure of the ever-moving claylike ground in the Czar-Holbrook Mine area. The vertical timbers are 25 centimeters by 25 centimeters in cross section and incapable of stopping the movement of the surrounding material. Some areas in these mines were unminable because of this difficult condition, 1904.

Figure 2.1.4: Subsidence crack with substantial displacement showing horizontal movement. By 1930 the crack had widened to the point that the bridge fell in to it. This was located to the south of the Holbrook Mine, 1913.

Figure 2.1.3: An example of the timber bulkhead used to keep heavy ground open as described in the 1904 *Bisbee Daily Review*. The heavy, 25 centimeters by 25 centimeters square timber is actually stacked on top of each other, several layers deep to support the opening. No space could be left open, as the clays would quickly extrude into the workings because of the intense pressure. Holbrook Mine, 1904.

The many huge cracks in Queen Hill were worrisome. More than a few in Bisbee truly believed the front of the hill would, someday, slide into the city nestled in the canyon below (Figure 2.1.5). By the late 1930s, the movement of the mountain had slowed and soon stopped as mining activity in the area diminished.

Mining of the New Southwest Orebody brought more cracks to the surface of the hills above Bisbee; most came with a huge collapse in 1933. Marginal subsidence cracks were reactivated and opened as the mining had removed the support. These cracks are in the upper portions of Hendricks Gulch. Figure 2.1.6 shows the interior of a small portion of the largely collapsed stope.

Figure 2.1.5: Subsidence cracks near the Holbrook Mine, as seen in 1937 showing substantial displacement along the cracks with the right side of each, down dropped, indicating movement toward Bisbee.

Figure 2.1.6: One of the authors in part of a collapsed cave over the New Southwest Orebody on the 6th level of the Southwest Mine. A few smallish stalactites can be seen on the ceiling directly above. This cave collapsed, in 1933 with subsequent falls occurring over the next few years. Cracks created by the subsidence associated with this collapse extend to the surface in the upper parts of Hendricks Gulch.

Chapter Two: Geology

Here, our focus will be on the geology and mineralogy as related to, and associated with, cave formation. Further information on the overall geology of the district can be found in Ransome, (1904a), Bonillas et al., (1916) as well as in Bryant and Metz, (1966).

About 180 million years ago, (Bryant & Metz, 1966) a thick sequence of Paleozoic Limestones and the underlying Precambrian metamorphic rocks were intruded by several pulses of magmatic activity interspersed with an episode of intense pyrite-silica mineralization between the two, main intrusive events. A large breccia mass was then emplaced more or less between the two porphyry intrusives. These events were followed by copper mineralization then, by lead/zinc deposition and a final phase of mineralization, which brought with it modest amounts of minor elements (Graeme, 1993).

The pyrite-silica phase mineralized the earlier of the intrusive units and invaded the limestones along faults and bedding planes, locally replacing large volumes of the limestone with pyrite and silica. This event formed numerous and typically isolated sulfide masses scattered throughout limestones. In all, more than 500 million tons of pyrite are estimated to have been deposited at Bisbee (Bryant & Metz, 1966). At least three subsequent mineralizing events occurred, which mineralized the later intrusive unit and followed the path of the pyrite into the limestones, replacing more of

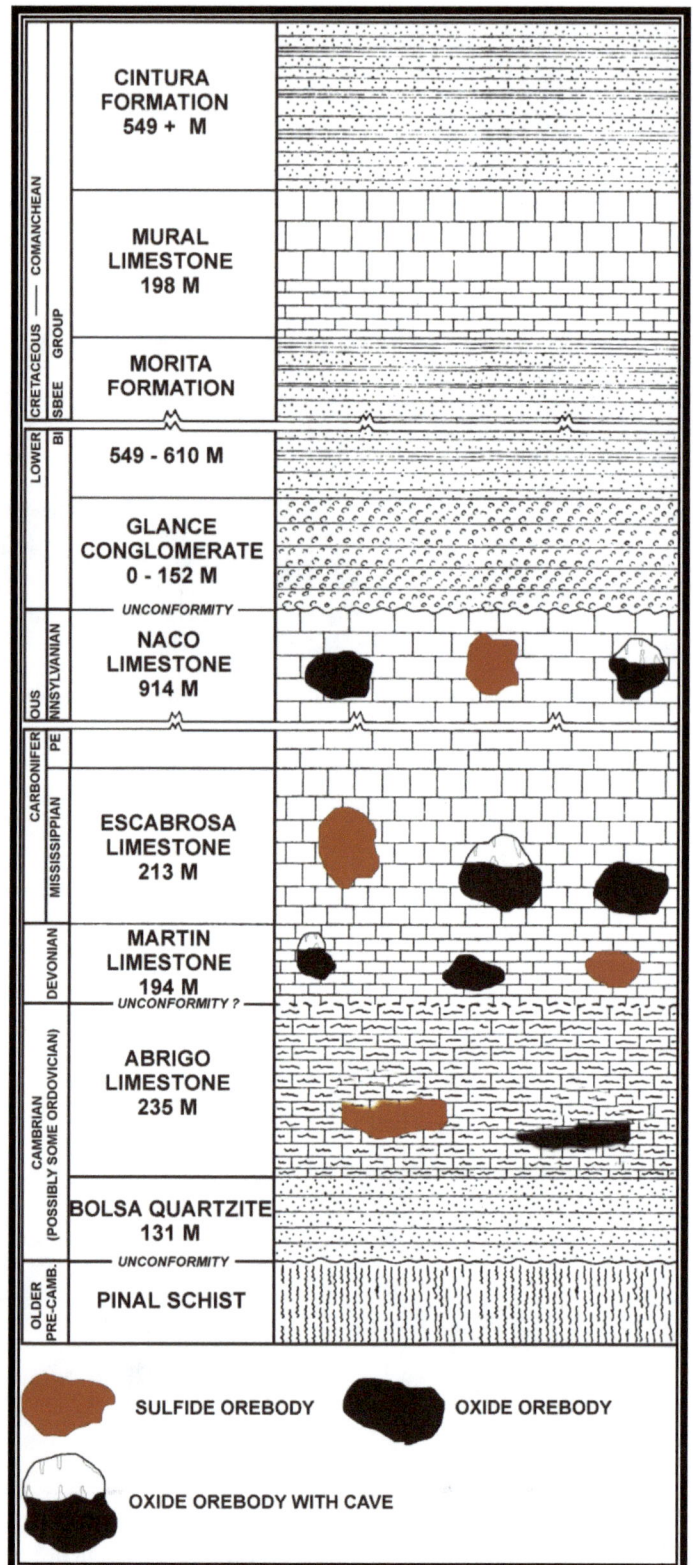

Figure 2.2.1: Generalized geologic section after Ransome, 1904a, Hogue and Wilson, 1950.

the carbonate rock and even some of the earlier massive pyrite (Bryant & Metz, 1966, Graeme, 1993).

Copper ore mineral deposition occurred along the footwall portion and peripheral to many of the earlier pyrite bodies and as smaller, replacement masses within the pyrite bodies (Bryant & Metz, 1966, Bonillas et al., 1916). Low grade, pyritic cores were a common feature of the copper sulfide deposits, as illustrated in Figure 2.2.2. Later lead and zinc mineralization followed a similar patter, but more typically formed more distal replacement type deposits in the same limestones. The final mineralizing episode was less intense and left a modest overprint of minor elements on the existing copper/copper iron/ iron sulfides. The elements from this event include vanadium, bismuth, tin, tellurium, gold and silver as well as a little copper, lead and zinc.

The result was many hundreds of typically isolated, sulfide masses, scattered throughout the limestones. Most of these replacement deposits were relatively large, ranging from several tens of thousands of tons to a few of more than a million tons of ore. The total deposited sulfide masses were much large when the non-ore sulfides are considered.

The replacement of the limestone by the sulfides was frequently a faithful process with some of the minor and many of the major features of the original limestone often preserved by the replacing sulfides. In one

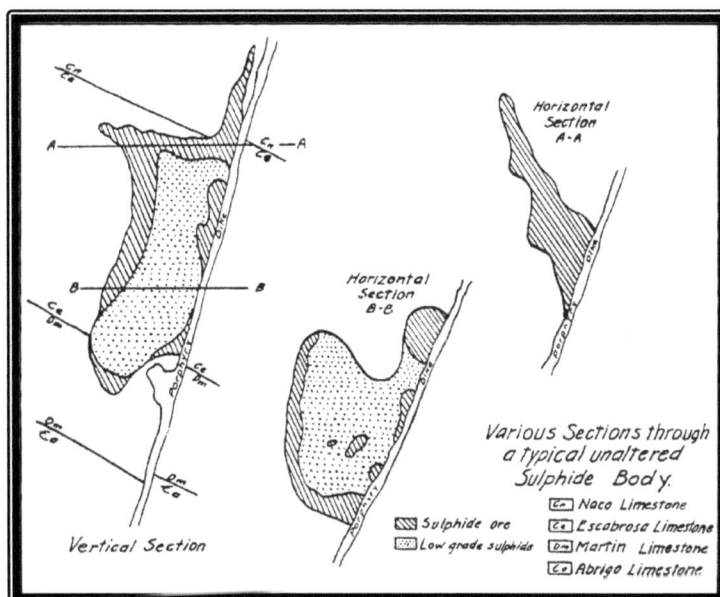

Figure 2.2.2: Views of a typical unaltered sulfide body (Trischka, 1938).

orebody mapped by Richard III on the 2700 level of the Campbell, a chert rich bed in the Martin Limestone could easily be traced from the unaltered limestone completely across the sulfide mass, some 20+ meters and into the seemingly fresh limestone on the other side of the orebody. On a micro-scale, in polished specimens, replacement features are commonly seen, both of the altered host rock and of the earlier pyrite by the ore minerals. This often includes features found in the now replaced, host limestone such a grain textures and even the occasional fossil fragment.

Features characteristic of open space filling by mineralizing fluids, such as large crystals of sulfide minerals, are conspicuously absent at Bisbee. Indeed, when Alan Criddle studied hundreds of sulfide specimens from the Campbell orebody, not once was a single crystal of any mineral seen without magnification (Alan Criddle, personal communication, 1992). This too strongly suggests that there were no preexisting caves to be filled with sulfide minerals during mineralization.

The carbonate-hosted orebodies at Bisbee occurred in a series of Paleozoic units that are more than 1,000 meters in total thickness. The Paleozoic limestone units are the Cambrian age Abrigo, Devonian period Martin, Mississippian Escabrosa and the Pennsylvanian-Permian Naco group. These deposits appear to radiate, spoke-like, to the south, from the Sacramento Stock complex, but in reality they followed a series of NE to SW trending structures, as shown in Figure 2.2.3.

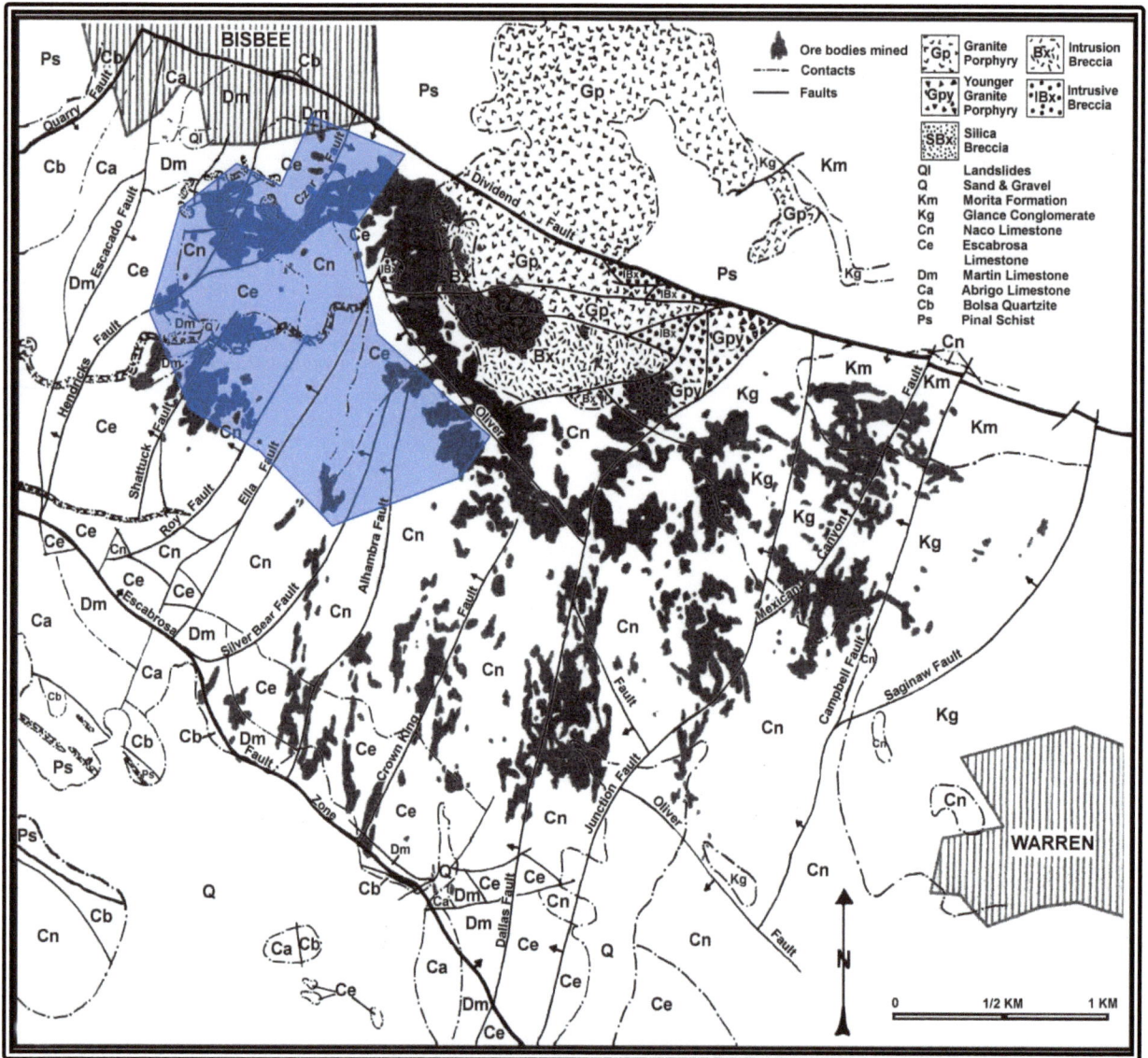

Figure 2.2.3: General surface geologic map with the ores mined by underground projected to the surface. The colored area on the above map indicates where oxidation caves are known to have occurred. After Graeme, (1981).

While ore was mined from all of these limestones, only the Abrigo, Martin and the Escabrosa Limestones contained any significant amounts. Of these units, caves were found only in the upper, thick-bedded portions of the Martin Formation, the lower, thicker bedded part of the Escabrosa Limestone and infrequently in the Naco Limestone. In all cases, the host units are relatively course-grained and pure limestones, although portions of the Martin are dolomitic. The Abrigo, a thin bedded, impure limestone was too weak to support cave development and simply broke up as the supporting rock below was removed by oxidation. This was also true for the lower, somewhat shaley, Martin Limestone.

Hydrothermal alteration and ground preparation prior to mineralization were important factors influencing ore deposition and subsequent cave development. While hydrothermal alteration was wide spread, it was typically modest in intensity except in the rocks immediately adjacent to the intrusive complex. Here, calci-silicate minerals replaced the limestone resulting in far less competent rock following oxidation. Supergene conditions tend to favor alteration to clays or claylike minerals in calci-silicate alter rocks. This does not favor cave development. Figure 2.2.3 shows that caves did not occur in the intensely hydrothermally altered area immediately adjacent to the intrusives. The whole of the contact zone around the Sacramento Stock complex was a topographic low, reflecting the poor rock quality because of the clays (Figure 2.2.4). In other areas, manganese had replaced up to 16% of the calcium in the limestone (Hewett & Rove, 1930). This reduced the solubility of the rock and impeded but did not stop cave development, as shrinkage of the oxides was the principle driver in this process.

Figure 2.2.4: Sacramento Hill area in 1904. Note the low nature of the surrounding areas in the foreground and to the right. This is a result of clay formation during supergene alteration of the hydrothermally altered limestones. The Holbrook Mine is at the lower left of the photo.

The forces associated with intrusion were more likely to deform the thin-bedded portions of the Martin Formation and all of the Abrigo Formation with some fracturing along bedding planes. In contrast, the thick-bedded units in the upper Martin Formation and the Escabrosa Limestone suffered substantial fracturing, both along bedding planes and vertically. While both deformational responses to these forces allowed access for hydrothermal solutions and sulfide mineral deposition, the fracturing in the thick-bedded units was more pervasive and extended well into the rock. The fracturing facilitated the development of preferred pathways for first, the mineralizing fluids and then the entry of groundwater, which caused the later supergene activity. Fractures also served as conduits for the low pH supergene fluids, which enhanced cave development.

Bryant and Metz (1966) have noted that the shapes of the sulfide deposits were different in the various limestone units. Those in the Abrigo were tabular and typically parallel to bedding with the width greater than the thickness. In the Martin Formation and Escabrosa Limestone, they were more "football" shaped with the vertical component being the greatest dimension. The shape of the orebody, influenced largely by pre-mineral fracturing and chemical characteristics of the host rock, would of course, predispose the subsequent shape of the cave that develops as the orebody is oxidized.

At some point before the Cretaceous Period (Bonillas, et al., 1916), the mineralized area surrounding Sacramento Hill was stripped by erosion and incised with a deep, steep walled canyon. Evidence for this paleocanyon can be found in the subsurface beneath Lowell, a small suburb of Bisbee and was clearly visible during the mining of the Lavender Pit. This exposure of the mineralized ground to oxygen and ground water in the Cretaceous Period triggered the first episode of supergene alteration, which affected the entire mineralized area. Supergene deposits in the Campbell Mine suggest that oxidation during this episode extend to approximately 250 meters below the paleo-surface.

 Later in the Cretaceous Period, the entire area dropped and was ultimately covered with thousands of meters of largely clastic sediments. Supergene alteration and associated oxidation would have been totally stopped, as the orebodies were now deeply buried and thus moved from near surface positions to great depth. Oxygen rich ground waters and bacteria are both necessary for oxidation and neither typically penetrate to any great depth. Final uplift occurred during the Pliocene Epoch. At this time, the region was cut by basin-and-range, normal faults and the entire Mule Mountains and mineralized district were tilted 15 degrees to the northeast. The uplift initiated erosional stripping of Cretaceous sediments from the western part of the range (Bryant & Metz, 1966). The removal of cover in the western portion of the district brought with it a second episode of localized supergene activity and the ultimate formation of the oxidation caves in this part of the mining district.

Chapter Three: Supergene Alteration as Related to Oxidation Cave Development

Supergene alteration oxidizes sulfide minerals exposed to near-surface environments, particularly with oxygenated water, which is a strong oxidizer. While it is a complex process, with several phases, a basic understanding of this alteration is important in the context of the role it plays in oxidation cave development. Only those aspects related to the formation of oxidation caves and the associated oxide ores are discussed below.

Pyrite, the common iron sulfide, is the most abundant sulfide mineral in the replacement deposits at Bisbee. When exposed to oxygen-rich groundwater in the near-surface environment, pyrite readily oxidizes, and begins a process, which forms ferric sulfate and ferric hydroxide in a sulfuric acid solution (Emmons, 1913), (Sato, 1960). Ferric sulfate is a very strong oxidizing agent and it attacks other sulfide minerals, liberating more ferric sulfate and other contained metals as soluble sulfates.

Figure 2.3.1: Massive gray pyrite oxidizing to goethite along a series of parallel faults, 7th level, Southwest Mine,

Figure 2.3.2: Goethite formed through the oxidation of pyrite with unoxidized pyrite core, Neptune Mine, Tunnel level, view - 5.5 centimeters, Harvard Mineral Museum collection.

Figure 2.3.3: Azurite veneer developed on oxidizing chalcopyrite, pyrite and bornite. Note the goethite derived from pyrite oxidation, as well as the iron contained in the copper bearing minerals, Gardner Mine, specimen - six centimeters.

124

The copper and copper/iron sulfide minerals such as chalcopyrite, chalcocite and bornite, which were often present in the Bisbee mineral deposits, oxidized readily, liberating into solution even more acid, ferric sulfate and mobilizing copper from within these minerals.

In the uppermost, oxygen rich environments, substantial portions of the oxidation generated acid are quickly neutralized and much of the iron is left in place as hydroxides and oxides which form the common goethite and with, dehydration, hematite, as shown in Figures 2.3.1 and 2.3.2. Typically, the copper liberated through oxidation remained with these residual iron oxides, redeposited as copper oxides or copper carbonates, not far from the original deposition site (Figure 2.3.3).

Figure 2.3.4: Acidic, iron-rich solutions, generated by the oxidation of pyrite, depositing iron hydroxide minerals on limestone with some goethite being deposited as darker patches, 2700 level, Junction Mine, vertical view 1.80 meters.

However, there is typically an excess of acid generated during pyrite oxidation. Substantial amounts of carbonate can be dissolved and mobilized under this excess acid condition (Bottrell, et al., 2000). Acidic, iron-rich solutions (see Figure 2.3.4) migrated downward into the host limestone along fractures and bedding planes, dissolving the buffering limestone, liberating substantial CO_2 and depositing the iron carbonate, siderite. Further oxidation and exposure to the acidic solutions in a high CO_2 environment, preferentially removes any remaining limestone from between the previously deposited siderite (Smith and Martell, 1976). This resulted in siderite boxwork forms, often lined with botryoidal siderite. Large volumes of siderite boxwork were deposited in this manner at Bisbee. For example, Trischka et al., (1929) noted a large mass of boxwork siderite, estimated to be some 30,000 cubic meters under a sulfide mass still in the oxidation process, which was indicated by the marked increase in temperature in the area where the siderite occurred. Sulfide oxidation is an exothermic process which can generate substantial heat, enough to ignite wooden timbers on occasion.

Figure 2.3.5: Siderite in boxwork form in the initial stages of altering to goethite, 1400 level, Sacramento Mine, specimen - ten centimeters, Harvard Mineral Museum specimen.

The estimated void spaces in this boxwork siderite were up to 40% (Trischka et al., 1929). Similar material is shown in Figure 2.3.5.

Figure 2.3.5: Siderite boxwork in place, 2300 level, Junction Mine, horizontal view - 2.1 meters, 1972.

Frequently, the voids within the siderite contained carbon dioxide gas when first found (Trischka et al., 1929). Bateman, et al., (1914) noted an instance in the Gardner Mine where high concentrations of carbon dioxide, liberated from voids in siderite, forced the withdrawal of miners for safety reasons.

Many other similarly large siderite masses such as described by Trischka et al., (1929) were found in Bisbee during mining and always beneath oxidizing sulfide masses. An

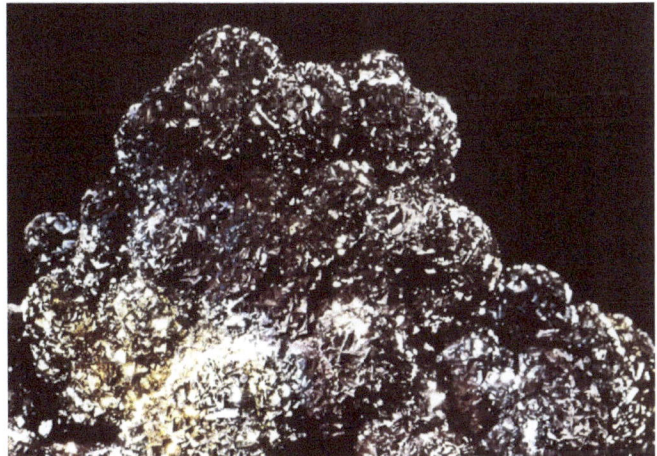

Figure 2.3.6: Uncommonly iridescent, botryoidal siderite, 2300 level, Junction Mine, specimen - 10.5 centimeters.

example of one such mass found on the 2300 level of the Junction Mine is shown in Figure 2.3.5. The portion illustrated is but a small part of a mass that extended more than 200 feet horizontally and in excess of 300 feet vertically.

It was intercepted by workings on the 2200 level, 2300 level, 2433 level and extended almost to the 2566 level. This siderite mass was an artifact of the pre-Cretaceous period, first phase of supergene activity. An estimated 20,000 tons of siderite were contained in this mass. A number of fine specimens were collected from here by the miners. An example of one such specimen is shown in Figure 2.3.6.

Trischka, et al., (1929) note that boxwork smithsonite (zinc carbonate) was found in similar environments in parts of Bisbee, something the authors have seen in the Holbrook Mine associated with oxidation cave development. An example of this material is in Figure 2.3.7.

Most of the copper liberated from the sulfides during oxidation remained with the residual iron oxides, redeposited as oxides and/or carbonates, as previously noted. This often resulted in a substantial increase in the copper content in the remaining material, as all of the sulfur and some of the iron had been removed through the oxidation process. This natural concentration process, called supergene enrichment, played a major role in forming the rich oxide orebodies found in and below the oxidation caves at Bisbee (Figure 2.3.8).

The oxidation of high-grade copper minerals such as bornite and chalcocite sulfate. These solutions attacked the limestone and deposited malachite and azurite in boxwork forms as they were quickly buffered by the limestone, as shown in Figure 2.3.9. In the iron oxides, high copper solutions, buffered by residual carbonate, redeposited the copper as linings of malachite and azurite within voids in the iron oxides. In these instances, the malachite or

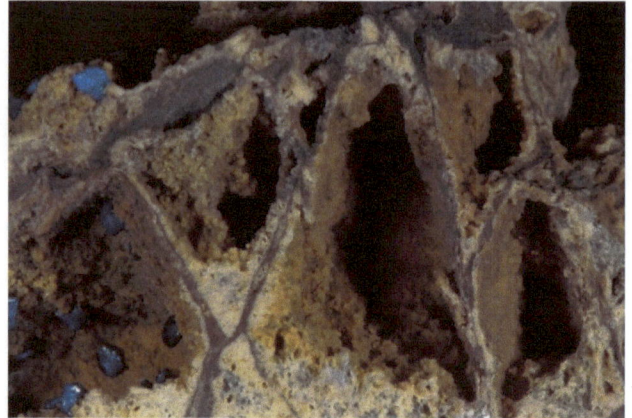

Figure 2.3.7: Smithsonite boxwork with minor azurite, Holbrook Extension, Lavender Pit Mine, view – seven centimeters.

Figure 2.3.8: Malachite and minor azurite with goethite resulting from the oxidation of iron and copper/iron sulfides. The copper content of this material has been enriched, and is several times higher than the unoxidized part of the same orebody. Note the abundant open space in the specimen, Sacramento Mine, view - nine centimeters.

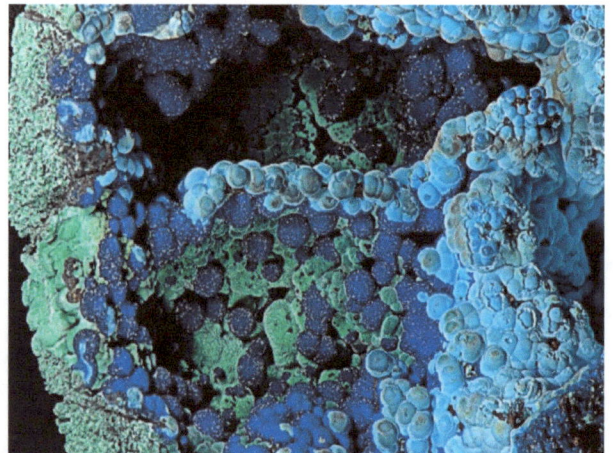

Figure 2.3.9: Malachite and azurite boxwork. The drusy, crystalline coating of the azurite spheres indicates period of minor, subsolution azurite deposition, Holbrook Mine, view - 15 centimeters.

azurite might be botryoidal, reflecting a subaerial environment (Figure 2.3.10) or as crystals, indicating a subsolution environment (Figure 2.3.11).

Near neutral, high copper solutions deposited the coatings and crust of malachite and azurite on the walls of the oxidation caves. Alternating bands of malachite and azurite spoke of subtle changes in the carbon dioxide levels in the cave atmosphere (figure 2.3.10).

The copper rich solutions also mixed with calcium carbonate waters to form calcite/aragonite speleothems tinted by the copper either as a substitution as discussed in Chapter Seven, or as co-deposited copper/calcium carbonates. In less acidic systems, high-grade copper minerals tend to oxidize in situ, altering directly to cuprite or, more commonly, directly to malachite as illustrated in Figures 2.3.12 and 2.3.13. This in situ alteration was responsible for substantial quantities of ore in some caves, such as the cave described on the 450 level of the Irish Mag Mine covered in Chapter 1.4.

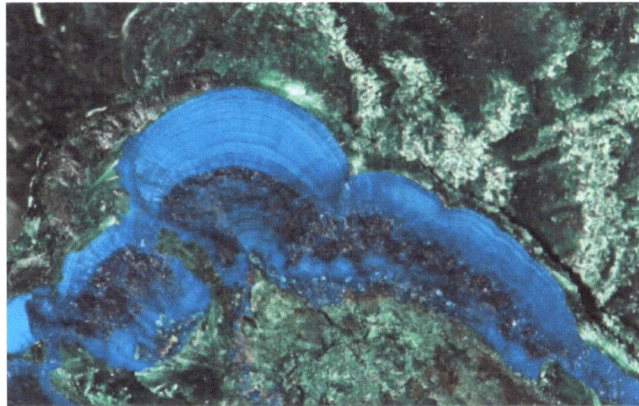

Figure 2.3.10: Alternating bands of malachite and azurite reflecting a changing depositional environment, Holbrook Mine, view - 18 centimeters.

Figure 2.3.11: Azurite with malachite pseudomorphs after azurite nearly filling a goethite, boxwork void, Gardner Mine, specimen - 14.7 centimeters.

Figure 2.3.12: Chalcocite altering directly to malachite along fractures. Note the openness of the malachite in the center of the specimen, Shattuck Mine, view - 7.7 centimeters.

Figure 2.3.13: Coarsely crystalline malachite derived from the direct, in situ alteration of chalcocite. Minor hematite and quartz are also present, Shattuck Mine, specimen - eight centimeters.

128

Every step of the oxidation process created open spaces, albeit small. Collectively these many small spaces made large openings, when the process was carried to completion and compaction had taken place with the collapse of the lower boxwork voids.

Chapter Four: Oxidation Cave Development

It is the considered opinion of the authors that the full and complete oxidation of the sulfide orebody is a necessary criterion for oxidation cave development. Siderite boxwork development was an intermediate step in oxidation cave formation. Ultimately, this siderite was altered to goethite by further exposure to acidic solutions as oxidation continued to generate acid (Trischka, et al., 1929).

Figure 2.4.1: Typical goethite boxwork with malachite and minor azurite, Copper Queen Mine, specimen - nine centimeters high.

Figure 2.4.2: Goethite that has been plastically deformed by the pressure of the overlying oxides, "B" level, Copper Queen Mine, view - 3.5 meters.

The resultant goethite generated by the complete oxidation of the sulfide orebody was often tens of meters thick, soft, clay-like and too weak to support the mass of goethite and rock above. Therefore, the boxwork structures in the lower areas collapsed and effectively transferred the many small, open spaces that were in the lower section of the orebody towards the top. This collapse and compaction of the supergene altered rock, which removed the support causing unstable conditions in the host limestone above the oxidation zone. The overlying limestones characteristically responded to this lack of support by a partial or total collapse and subsiding as well as dilation of the limestone beds, creating open spaces and oxidation cave development. Subsidence occurred gradually as the supporting minerals were removed. Combinations of all of these features were common above thoroughly oxidized orebodies. Breakdown, or collapse as Wisser (1927) called it, was an important secondary mechanism for oxidation cave formation. Consequently, boulders of the host limestone, often partially altered by supergene activity, were typically abundant in the cave bottoms as were boulders of goethite and hematite, all having fallen from the walls and ceiling of the cave during development. Occasionally, chunks of ore minerals, such as malachite, were in the breakdown rubble.

Figure 2.4.3 shows chunks of malachite with angular fragments of unaltered limestone and pieces of goethite from the rubble of a cave floor. All of these materials had fallen from the cave ceiling or walls as a part of the breakdown process during cave development and are now cemented by calcite. The depth of the rubble often exceeded several meters in the larger openings. In some of the caves, later, thick travertine deposition frequently obscured the breakdown deposit in the cave bottoms.

Beneath the breakdown and in the oxidation generated iron oxides were abundant supergene copper minerals, as illustrated in Figure 2.4.4. Malachite and azurite were the most common and occurred as large masses and linings in the voids between boulders or lining the surviving goethite boxwork. On rare occasion, the rubble in the cave bottom was largely composed of goethite with

malachite and azurite, often in large pieces, and little to no limestone present. These occurrences were extraordinary and the sources of incredible mineral specimens. Obviously, these areas were rich ore deposits as well. In a few, rare instances the cave bottoms were not mined because of low copper content. The authors have found abundant, late stage, supergene, copper minerals including boulders of spongy malachite approaching a meter in size in several un-mined cave bottoms, an example of which can be seen in Figure 2.4.4.

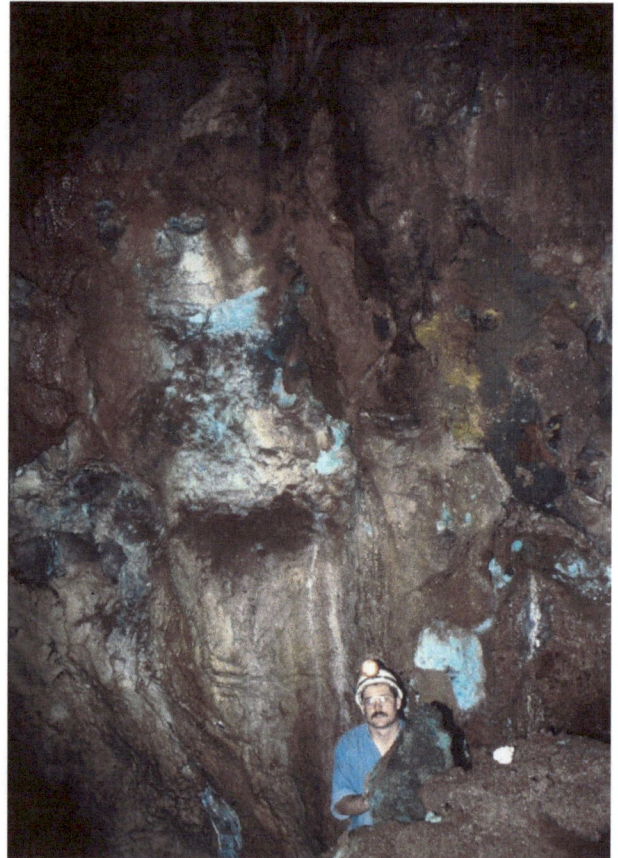

Figure 2.4.4: One of the authors with a 20-kilogram boulder of malachite found in a cave floor, 100 level, Holbrook Mine.

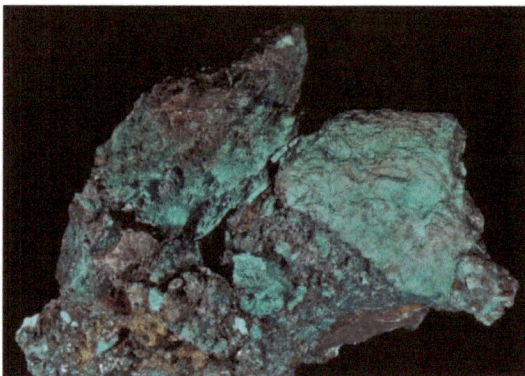

Figure 2.4.3: Malachite chunks with unaltered limestone fragments and goethite cemented with calcite from a cave floor. These represent breakdown materials that fell from the cave ceiling or walls, 100 level, Holbrook Mine, specimen – 60 centimeters wide.

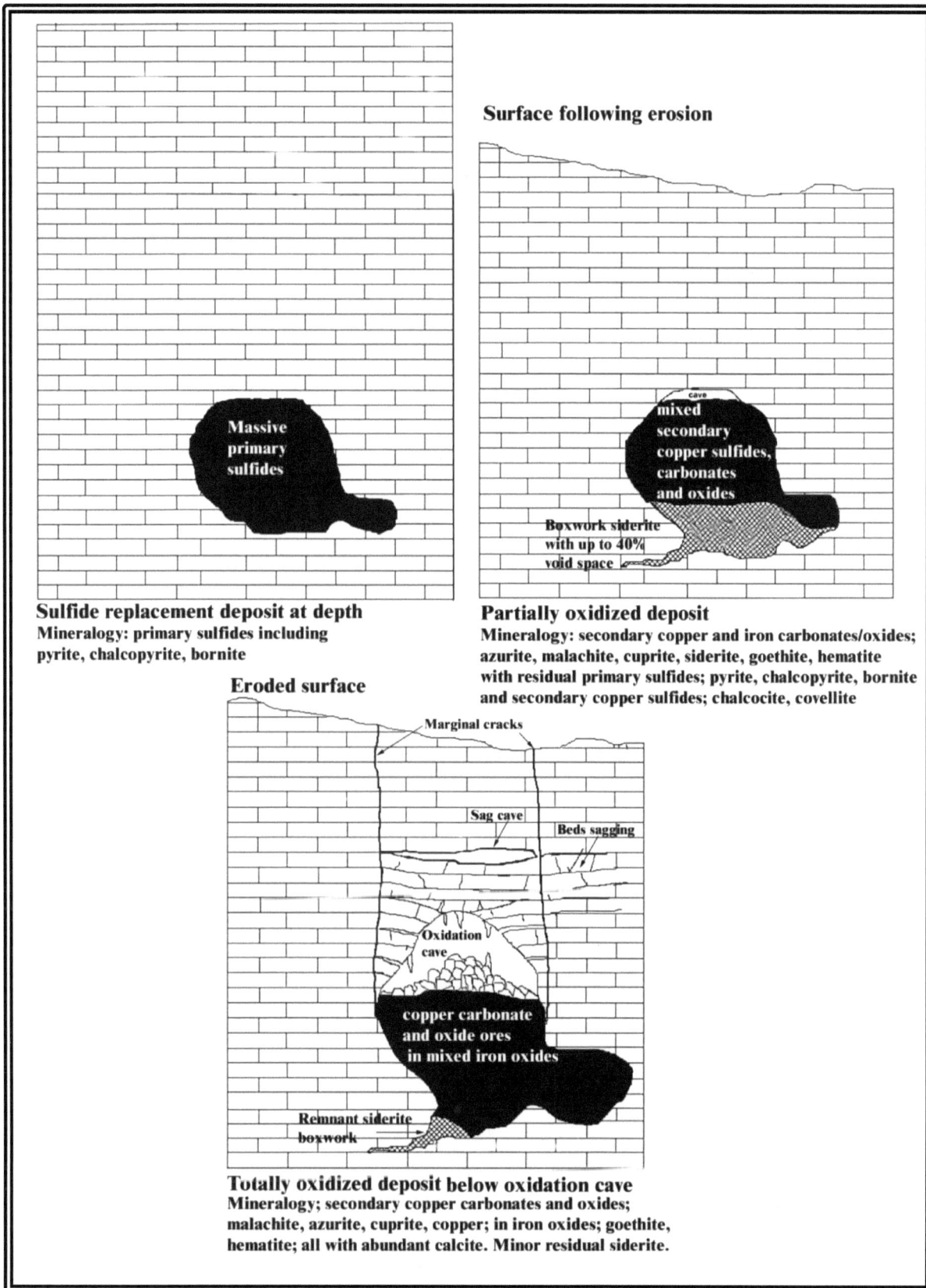

Surface following erosion

Massive primary sulfides

Sulfide replacement deposit at depth
Mineralogy: primary sulfides including
pyrite, chalcopyrite, bornite

cave

mixed secondary copper sulfides, carbonates and oxides

Boxwork siderite with up to 40% void space

Partially oxidized deposit
Mineralogy: secondary copper and iron carbonates/oxides;
azurite, malachite, cuprite, siderite, goethite, hematite
with residual primary sulfides; pyrite, chalcopyrite, bornite
and secondary copper sulfides; chalcocite, covellite

Eroded surface

Marginal cracks

Sag cave

Beds sagging

Oxidation cave

copper carbonate and oxide ores in mixed iron oxides

Remnant siderite boxwork

Totally oxidized deposit below oxidation cave
Mineralogy; secondary copper carbonates and oxides;
malachite, azurite, cuprite, copper; in iron oxides; goethite,
hematite; all with abundant calcite. Minor residual siderite.

Figure 2.4.5 Schematic representation of the development of an oxidation cave.

132

A second type of cave formed by oxidation was what the miners referred to as "sag caves." Wisser (1927) also employed this appropriate term. These openings formed along bedding plains in the limestone above and adjacent to the oxidation caves. As the support below was removed by oxidation subsidence, openings developed through the dilation of the beds. It was uncommon for these types of caves to contain either iron oxides or ore minerals as they were well above the orebody. Also, the caves formed by dilation were typically small. Usually, the dilation of the beds did not exceeded two meters in height. However, they could be up to 20 meters in length. All of these features are illustrated in Figure 2.4.5, which diagrammatically illustrates the stages in the development of an oxidation cave over an orebody as it progressively, oxidizes and the formation of sag caves in the overlying limestones.

Given this strong correlation between oxidation of sulfide ores and subsidence with cave formation, a great deal of effort was spent looking for those cave related features, which would indicate the possible presence of ore (Douglas, 1900, Wisser, 1927). Features such as caves, subsidence generated marginal cracks, sag caves and the presence of limestone broken because of subsidence below (Figure 2.4.6), were all indicators that suggested ore might be some place below. Siderite or massive goethite, which had replaced siderite, would indicate to the miners that they must look upward in their search for ore.

Figure 2.4.6: A crosscut cut through blocky, chaotic boulders of broken limestone showing dilation and breakage along bedding planes, subsequently cemented by minor calcite. This is located more than 100 feet above an oxidation cave and the collapse of the beds was caused by subsidence due to the removal of support below because of oxidation and the formation of a cave. A feature such as this was an important guide to possible ore somewhere below. This is 86 Crosscut on the 6th level of the Southwest Mine and mining through this broken rock would have been both difficult and dangerous as large blocks of limestone fell from the sides and back (top). Maximum width - 3.5 meters.

Chapter Five: Oxidation Cave Features

There are a several characteristics common to all oxidation caves studied. All are illustrated in part three of Figure 2.4.5. These features are:

- Direct association with large amounts of sulfide oxidation products such as supergene goethite and hematite
- They are isolated, single chamber features not connected to or related to other caves, with rare exceptions were a very large oxidized orebody may host more than a single chamber
- All are relatively close to the current surface, in the western part of the district only, where two episodes of supergene activity occurred
- Subsidence related cracks along the margins of the oxidation cave, extending vertically for great distances

The single most important feature is the association of oxidation caves with thoroughly oxidized mineral deposits. All oxidation caves occur above and/or partially within masses of typically mixed oxide minerals, the more common of which are goethite and hematite. At Bisbee, substantial amounts of copper ore minerals often occurred within the iron oxides associated with the oxidation caves. The oxide masses associated with these caves were usually quite large, often 30 meters or more thick. The lower portions of the oxide mass beneath the cave were invariably quite compact as a result of the weight of the overlying material coupled with the soft, often earthy and clay-like nature of these oxides. In the upper most portions of the iron oxide mass, voids including boxwork structures did survive. These boxwork voids were often the depositional sites for malachite, azurite and other copper minerals as were the voids in the associated rubble pile formed in the cave bottom by boulders of oxide minerals and limestone derived from breakdown.

Oxidation caves are always very limited in extent, occurring as isolated, single-chamber openings whose size and shape was a direct function of the size, shape and orientation of the original sulfide deposit as well as the nature of the host rock. In the very few caves where more than one chamber appeared to exist, it was typically an expression of localized breakdown accumulation nearly filling the opening; travertine or speleothem deposition forming a partition-like wall or by variable dilation along several bedding plains in the host rock.

Very large orebodies frequently had more than one cave develop over the oxide ores. The large cave on the 300 level of the Shattuck Mine was, in reality two oxidation caves which formed over the opposite extreme of a mixed copper/lead orebody. The two caves were separate on the 200 level and above, but merged on the 300 level, where both caves were at their largest (Figure 2.5.1).

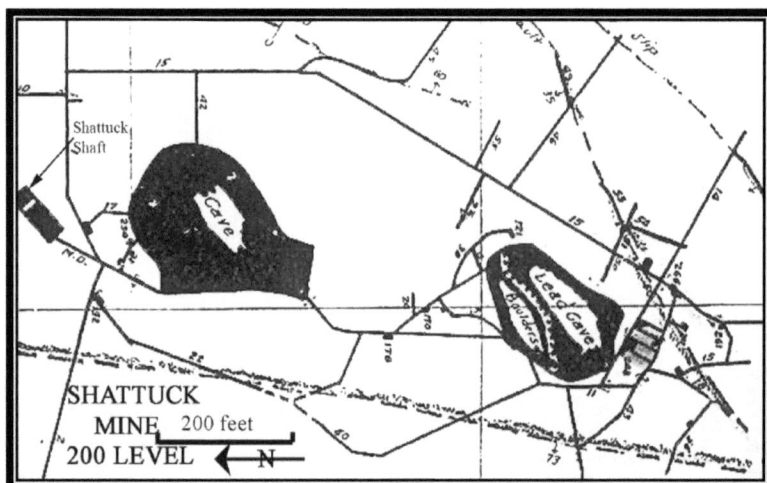

Figure 2.5.1: Left, map of the mine workings on the 200 level of the Shattuck Mine, showing the outline of the two, separate caves at this elevation. By less than ten meters below this elevation, they had joined, forming a single, large cave.

The vast majority of these caves were small, seldom more than 30 meters across and rarely more than ten meters high. Indeed, many were even smaller, but there were notable exceptions, several of which were discussed in section one.

All of the oxidation caves in Bisbee occur within 300 meters of the present surface. This is not too surprising since sulfide oxidation, which plays the major role in forming these caves, is limited by the depth to which oxygen-rich water penetrates below the surface.

While minor sulfide oxidation has been recognized at Bisbee to depths exceeding 800 meters along major geologic structures and intense oxidation was also common to a depth of 600 meters. This deeper supergene oxidation is largely an artifact of the first episode of activity that occurred during pre-Cretaceous times and is only so deep because of subsequent burial by Cretaceous sediments. Because the first phase of supergene activity did not totally oxidize the orebodies it affected, there are no oxidation caves associated with orebodies that experienced just this first phase of oxidation. There can be little doubt however, that it did contribute to their development in the western part of the district.

Oxidation caves are, without exception, restricted to the western part of the mineralized area, that portion of the district, which experienced two, distinct phases of supergene activity (See Figure 2.2.3). It was the second phase of supergene activity that completed the oxidation of the orebodies and caused the formation of the caves.

The marginal subsidence cracks are unique to oxidation caves. These are, as the name suggest, cracks which form around the margins of the cave as the overlying limestone plug drops in response to the removal of support from below caused by oxidation shrinkage. This feature, alone, is proof that the caves did not exist before mineralization took place or these cracks would have been conduits for hypogene mineralizing fluids. This was not the case, however. To be sure supergene solutions followed these features. Wisser, (1927) observed that marginal cracks were

135

commonly lined with, if not completely filled by travertine type calcite, indicating openness to solution passage. This is shown in figure 2.5.3 below.

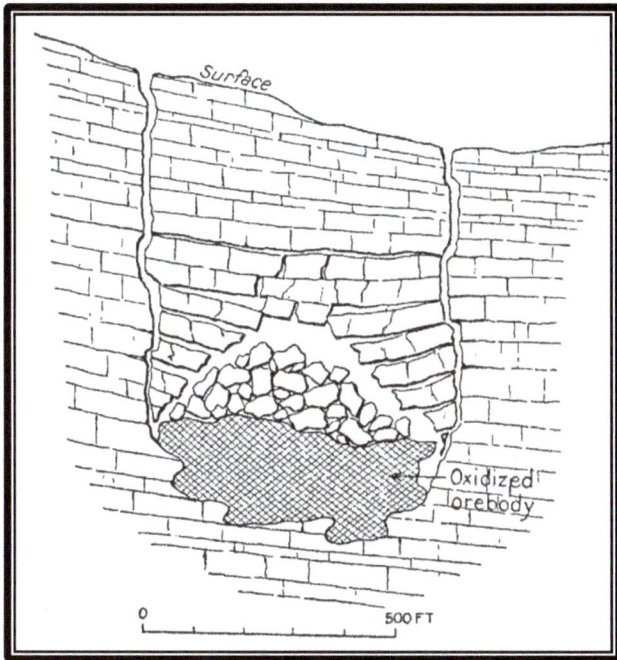

Figure 2.5.2: Diagram showing marginal crack along the cave boundaries (Wisser 1927)

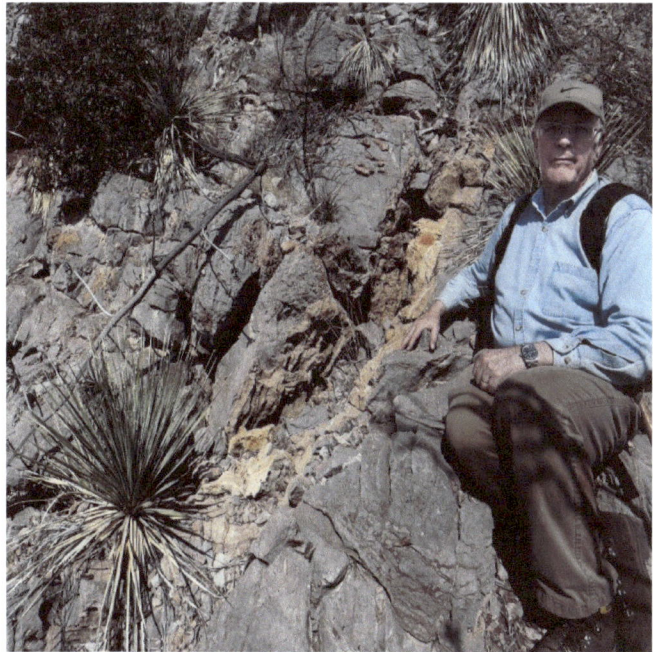

Figure 2.5.3: One of the authors next to a calcite filled marginal crack on the east slope of Queen Hill. Note: the related cave/orebody was at least 100 meters below this point on the surface.

Chapter Six: The Absence of Other Cave Types at Bisbee

The 25-kilometer long Mule Mountain Range, in which Bisbee is situated, is, in large part composed of limestone representing various Paleozoic and Cenozoic units. However, in spite of the abundance of limestone, there are but two known, small natural caves, both well outside the mineralized area and both in Naco Limestone. One lies 14 kilometers to the northwest of the city of Bisbee while the other is just over ten kilometers due west.

Localized modern karst development has been recognized by the authors south of the Escabrosa Fault and outside of the mineralized area. The Naco Limestone has a few, very localized characteristics typical of karstic development, including features suggesting collapsed caves (Figure 2.6.1). Neither of the two exploration mines of any size in this area, the Arizona Bisbee Copper Mine and the Bisbee West Mine, encountered any caves or solution cavities of note. However, it is more than a kilometer from either mine to the above noted features.

Figure 2.6.1: Looking southwest from near Mount Ballard toward the San Pedro Valley. The Bisbee West Mine is in the upper center of the photo. The Naco Limestone formations in the lower to center right show substantial karst development, 1902, (Photo from Ransome, 1904a).

Minor karstic enlargement of faults in the limestone is a common occurrence at Bisbee, as would be expected. However, this was highly localized in nature and never extensive or large with the two largest described below. Miners referred to these openings as "watercourses" as they frequently would contain modest amounts of water when found. Most drained soon after being hit by mine workings and usually remained dry. In the mine workings nearest to the surface, rainwater would quickly flow from these openings following a significant rainfall event, but they would soon be dry again. There is no recognized connection between these openings and the oxidation cave development.

Within the five kilometer by three kilometer mineralized area, there is an absence of non-oxidation related caves. It would be reasonable to expect that if there were caves, they would have been found, given the immense amount of underground development (>3,500 kilometers of workings) and diamond drilling that took place (>4,000 kilometers of exploration drilling). In all, five small openings not directly associated with masses of supergene minerals are known to have been hit during mining and exploration drilling in the mineralized area. Most were little more than barren, apparently post-mineral hydrothermally developed openings.

Three of these openings have been visited by coauthor, Richard Graeme III. The caves visited were on the 2200 level of the Junction Mine, 2700 level of the Campbell Mine, while the third occurred on the 3100 level of the Denn Mine. A brief description of those caves follows.

The 2200 level cave was perhaps 20 meters along strike and an easy 15 meters wide and, while backfilled to a point even with the 2200 level; it was still at least ten meters high. This cave had formed in the hanging wall of the Junction Fault, not too far from the Denn Sideline. It appeared to be also strongly influenced by bedding. The cave was hosted in a very friable, slightly altered Martin Limestone, which was a light reddish color and most visible surfaces were angular, possible due to post-formation breakdown. Notably, it was dry and absolutely devoid of calcite of any form. No other minerals of any kind were present in either the cave or the adjacent hosting limestone. The opening was, most probably, hydrothermal in origin.

On the 2700 level, the cave visited was well east of any known ore on that level. It was a structurally controlled opening at the edge of a series of closely spaced, near vertical faults in fresh limestone. The single open space of note contained no calcite and was but 15 meters along strike and no more than several meters wide at its widest point. The vertical extent was undeterminable when visited in 1962, as nearly 2,000 liters per minute of water was flowing from the shear zone, much of which was cascading through the cave. This system was the source of Bisbee's only mine flooding event, in spite of the substantial water found in the mines from the very beginning.

This shear zone/cave was, unfortunately, near parallel to the exploration crosscut, thus drilling well ahead of the area to be blasted did not detect the presence of water, so close to the course of the working. Following a blast at shift change in August 1941, the thin layer of rock in the crosscut wall broke under the pressure of the water. The ensuing water flow exceeded 19,000 liters per minute (Mills, 1958). The massive steel water door on the 2700 had been closed and sealed before the blast, but the water door on the 2566 level, some 40 meters above, had been left open. The heavy flow quickly filled the 2700 level behind the water door and then filled a vertical connection between the levels and behind the water doors on both levels, to flow through the partially open water door on the 2566 level. The twin, 7,570 liter per minute (2,000 gallons per minute) pumps on the 2700 level Campbell were quickly overwhelmed by the flow. Soon, the pumps on 2700 level Junction Mine, just over a kilometer away by underground connection, were underwater. Both mines were filled to about ten meters below the 2433 level (Mills, 1958). It took more than a year of sustained pumping and bailing to recover the 2700 level. This was the greatest flow of water ever hit during mining at Bisbee.

The extent of this system fracture was never determined. Extensive drilling on both the 2700 and 2566 levels cut the shear zone, but never encountered any open spaces similar to that seen on the 2700 level. Thus, it may have been just a wide, water filled structural zone with a few solution enlarged openings and perhaps even minor post-ore karst development as post-mineralization movement along the fault has been determined by Phelps Dodge geologist in other parts of the mine. However, this is speculative at best, as solution enlarged areas in fault zones were common. Interestingly, while there were no speleothems of any type in the cave or along the surfaces of the shear zone, several centimeters of tan to cream-colored post-mining calcite were deposited on the mine floor, beneath the fast flowing water in a bit less than 20years.

On the 3100 level there is a 40 meter long opening, which was bedding controlled, extending some 20 meters along the bedding plan and was never more than two meters in true height. This cave was totally devoid of any calcite forms, but small clusters of some post-mining sulfates, possibly epsomite (magnesium sulfate), were in scattered patches along the walls. It was water filled when first found in the early 1950s however, the cave was dry in 1960, indicating a lack of hydraulic connection. Even though the cave was located well east of any known ore on that level, nearby structures were known conduits for mineralization on other levels in other parts of the mine. Thus, it would seem possible that this cave may have been hydrothermal in origin.

Outside of the economically productive area, a single, small cave was found in the Bisbee Queen Mine, which is located more than three kilometers southeast of any known ore. This cave occurred at the contact between a slightly mineralized, hydrothermal silica breccia and the hosting limestone (Figure 2.6.2). While it contained a few normal speleothems, it was almost certainly hydrothermal in origin. This is strongly suggested by the morphology of the early calcite crystal forms which were deposited directly on the limestone and which were pre-speleothem development in age as they are partially covered by cave type calcite.

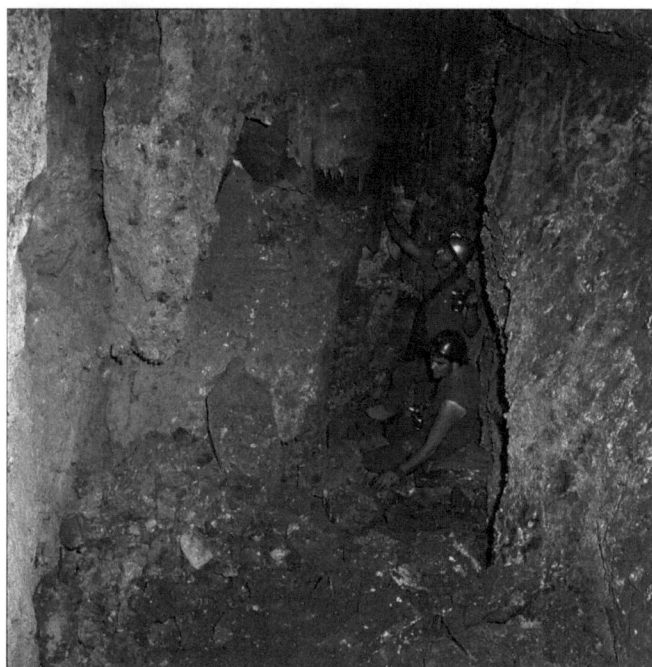

Figure 2.6.2: Cave like opening on the 100 level of the Bisbee Queen Mine in 1967. The right wall is a silica breccia, while the left is limestone. The broken rock in the bottom of the photo is from an exploration working sunk downward for most of 20 meters searching for oxide copper ores that did not exist.

The miners must have been ecstatic when they first hit this small cave, as it could easily have been confused with those that formed along marginal cracks. They mined downward for nearly 20 meters seeking ore, as was the practices at Bisbee, but found nothing save a few, impure iron oxides derived from the weak mineralization in the silica breccia. The opening extended more than three meters above, but only a few meters below the100 level.

Obvious hydrothermal voids were found in most of the mines, often near sulfide deposits, and many contained substantial amounts of crystalline calcite (DeWilde, 1915). These types of openings, while not uncommon, were never abundant. Most were small and in a few cases, goethite was found as a coating of the calcite (Figure 2.6.3).

The calcite lining these openings was obviously hydrothermal in origin as indicated by crystal morphology and too, because much of it has a strong, red-orange fluorescence. None of these hydrothermally developed openings contained sulfide minerals, even in minor amounts. This suggests that they may have developed post sulfide mineralization.

Figure 2.6.3: Calcite with a light coating of goethite lining a hydrothermally developed void, 2700 level, Campbell Mine. Photo taken in 1973.

The most extensive of these hydrothermal enlarged features seen by any the authors was hit in early 1947, just below the 2433 level and largely accessed through a stope from the 2566 level of the Campbell Mine. It occurred in hard barren limestone some five meters from, and roughly parallel to, an important lead/zinc orebody. This irregularly shaped opening extended along the trend of the bedding plane for more than 20 meters and ranged from 1.5 meters to three meters in width and was always less than two meters in

Figure 2.6.4: Hydrothermal calcite with phantom growth lines, 2566 level, Campbell Mine, view - five centimeters.

true height. All sides of the opening were covered with one centimeter size, tabular, pseudohexagonal calcite crystals (Figure 2.6.4). Most were lightly coated by goethite and a later, partial calcite overgrowth on the goethite (Figure 2.6.5). Both generations of calcite are highly fluorescent Figure 2.6.6). Many hundreds of interesting specimens were collected by the miners from here over the years.

Figure 2.6.5: Pseudohexagonal calcite crystals lightly coated with goethite, 2566 level, Campbell Mine, specimen - 17 centimeters.

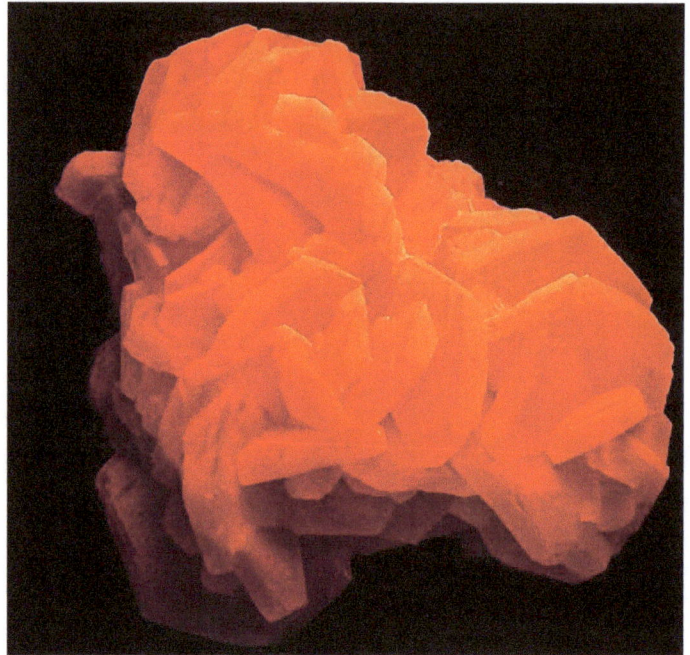

Figure 2.6.: Hydrothermal calcite response to shortwave UV light, 2566 level Campbell Mine, specimen - 6.2 centimeters.

Well east of the Campbell Shaft on the 2200 level was another large, calcite lined, hydrothermal opening. Here, along the Saginaw Fault was a meter-wide, opening which continued for 20 meters in length and six to ten meters in height along the steep dip of the fault. It was lined with amber to light tan calcite scalenohedrons that were from seven to ten centimeters long. The crystals had been deposited directly on hard, fresh limestone.

In a nearby hydrothermally developed opening, quartz cast pseudomorphs of similarly sized scalenohedral crystals were found. The quartz was in turn coated by centimeter sized, modified scalenohedral calcite crystals which had been lightly coated by goethite.

Figure 2.6.7: Lightly iron tinted calcite from a relic, hydrothermally developed opening in the oxidized area. This form of calcite had totally replaced the earlier, hydrothermal calcite, 5th level, Southwest Mine, specimen - 18.5 centimeters.

Hydrothermal openings in the oxide zones were typically unrecognized because of the intense oxidation overprint, as well as the abundance of oxidation developed open spaces. Also, most of the original hydrothermal calcite had been deeply etched if not completely dissolved away and new calcite deposited. The opening sides were usually covered by oxide zone type crystalline calcite during the oxidation process. An example of this type of calcite is shown in Figure 2.6.7. Often, cave type calcite was deposited in these openings as late stage overgrowths of the original hydrothermal calcite, if any remained.

This distinct lack of other caves in the mineralized area argues against the concept that the ores at Bisbee might have been deposited in pre-existing caves associated with a paleokarst system. While it is not rare for mineral deposits to occur in paleokarst environments, this was not the case at Bisbee. There are no recognized paleokarst features in the entire Bisbee area.

Breccias are commonly associated paleokarst system ore deposits as noted by Quinlan (1974), Walker (1928) and others. These paleokarst breccias are typically relics of the normal breakdown which occurred during cave formation.

While there are a number of breccias associated with the ores at Bisbee, there is no compelling reason to suggest these breccias are of this type. Dikelike, multi-lithology, intrusive breccias are associated with the ores district wide. These breccias contain rounded to well rounded fragments of all pre-Cretaceous rock units as well as sulfides (Bryant & Metz, 1966). Figure 2.6.8 illustrates this type of breccia.

Figure 2.6.8: Multi-lithology breccia adjacent to and above oxidation caves associated with the New Southwest Orebody. Most of the breccia fragments have been replaced by silica. The ground mass is largely silica and specular hematite, 7th level, Southwest Mine, horizontal view - 2.8 meters.

Figure 2.6.9: "Silica Breccia" composed of siliceous, altered limestone with silica fragments and minor goethite in a groundmass of quartz and specular hematite, Uncle Sam Mine, specimen - nine centimeters.

In the area where the oxidation caves occur, most breccias are composed of silica replacements of angular limestone fragments in a quartz/specular hematite matrix of definite hydrothermal origin (see Figures 2.6.8 and 2.6.9). These breccias are associated with only a small percentage of the caves. Bonillas, et al., (1916) and Trischka (1928), (1932) noted the connection between these breccias and intrusives at depth (see Figure 2.6.10). In a breccia associated with a cave in the Southwest Mine area, the authors have found relic fossils in several breccia fragments which would indicate an upward movement of these units, not downward as collapse or filling would bring. Further, there is little similarity with these breccias to the karst related breccias as discussed by Quinlan (1974), Dźulyński and Sass-Gustkiewiez (1989), Bosák (1989), as well as Walker (1928).

That ores were definitely not deposited in pre-existing caves or openings is also demonstrated by the fact that hundreds of ore bodies, both hypogene and supergene were mined that were not associated with any type of opening, hydrothermal or supergene developed.

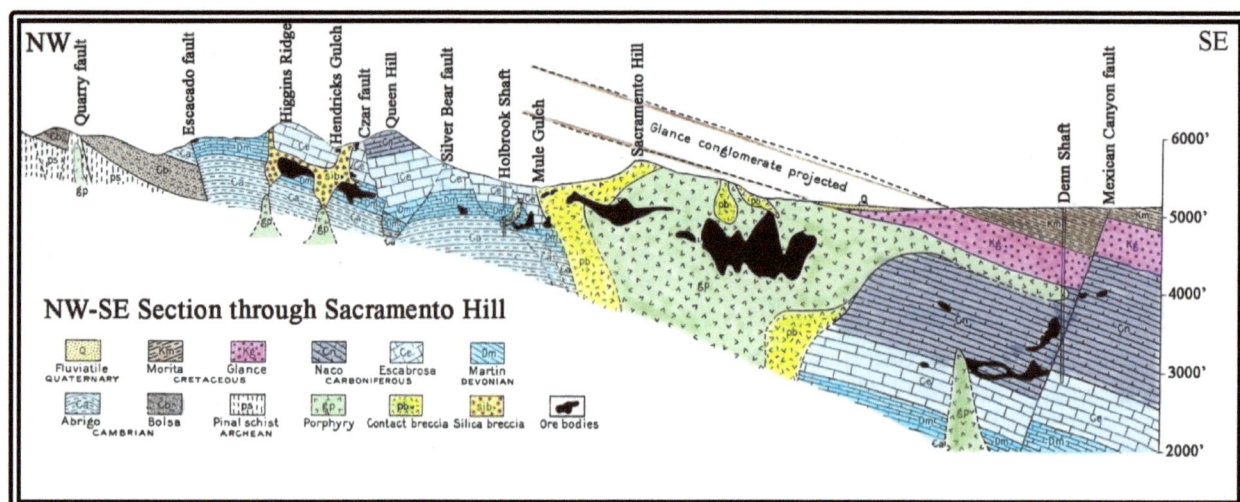

Figure 2.6.10: A NW-SE geologic cross section through Sacramento Hill. Note the silica breccias in the Higgins Ridge/Hendricks Gulch area and their connection to porphyry intrusives at depth. After Bonillas, et al., (1916).

There is a notable absence of sulfide mineral crystals at Bisbee. Large sulfide mineral crystals are a feature typical of sulfide deposits emplaced in preexisting voids. This lack of sulfide mineral crystals is true throughout the whole of the mineralized area including the western portion.

In our view, all of these factors are further indications that the sulfides were not deposited in pre-existing open spaces and that there is little chance the caves associated with the oxide ores were developed by mechanisms other than sulfide oxidation.

Chapter Seven: Speleothems Types Recognized at Bisbee

The oxidation caves at Bisbee were typically well decorated and displayed a wide variety of speleothems. As would be expected, stalactites and stalagmites were the more abundant. In an effort to more completely describe Bisbee's caves, the recognized speleothem types are discussed below. It is important to note that the names given to various speleothems by the caving community do not always correlate with the standard mineralogical or geological terms and that they are often fanciful, but quite descriptive and widely accepted. Therefore, the commonly accepted speleothem terminology from Hill and Forti (1997) are employed to describe the forms the authors and others have recognized at Bisbee. Where possible, photographs of the speleothem types are shown following the descriptions.

Anthodites: An impressive cluster of calcite anthodites was noted in photographs of the Shattuck Cave as seen in Figure 1.4.8. Also, other historic photographs show anthodites in different caves. Radiating clusters of light blue-green aragonite anthodites with calcite overgrowth were noted in a single locality by the authors in the Southwest Mine (Figure 2.7.1). Hill (1981) reports a second occurrence of aragonite anthodites in this mine in the "Higgins Mine Cave," which is actually in the Southwest Mine, is discussed in Chapter 1.5.

Boxwork: Because oxidation caves are a product of supergene oxidation, boxwork forms were abundant in a number of the caves. The more common minerals that formed the boxwork include goethite, calcite, siderite and smithsonite. It did occur on cave walls, as illustrated in Figure 2.7.2. Boxwork was far more abundant in the material in the cave bottom, often extending for some depth into the oxides below the cave.

Cave clouds: Cave clouds at Bisbee formed due to a localized perched water environment and have been found in a few small caves in the Southwest and Shattuck mines. The example illustrated in Figure 2.7.3 is from a small, isolated oxidation cave that collapsed after the ore below had been removed. The calcite-cemented breakdown is visible in the bottom of the photo.

Cave cups: Monocrystalline cave cups have been found in several localities as pool deposits that formed at the water – air interface. These were not common at any of the localities and were white to cream in color. An example is shown in Figure 2.7.4.

Coatings: Most oxidation caves contained thin coatings of one mineral or another. Calcite was the more common, followed by aragonite, but other minerals such as malachite, azurite and chrysocolla also occurred as coatings. Examples of this can be seen in Figure 2.7.5 and Figure 2.7.6, as well as other photographs of the cave on the 100 level Holbrook Mine cave discussed in Chapter 1.7.

Columns: This is a well-known form where stalactites join with stalagmites below. Columns were found in many of the caves and in a few instance occurred as strikingly beautiful masses in excess of ten meters high. Figure 2.7.7 shows a number of iron colored columns. As with stalactites and stalagmites, coloration by iron and copper of the columns was locally common in these caves.

Coralloids: These forms are often represented by the "popcorn" growths (Figure 2.7.8) that were found in the vast majority of the oxidation caves. Popcorn usually occurred as a late stage deposition on many of the other speleothems. It was also common for crystalline calcite or aragonite to be partially or completely cover with popcorn, reflecting the late phase deposition.

Typically white, the popcorn was occasionally found cream to a light tan and in one case deep red brown popcorn was observed. In a few rare instances, the popcorn was tinted a light green by copper. Popcorn knobs up to eight centimeters in length with microcrystalline surfaces have been recognized in one cave, an example of which is illustrated in Figure 1.7.6. Aragonite frostwork was often found on popcorn calcite as can be seen in Figure 2.7.8.

Crust: Crusts were abundant in the majority of caves with calcite being the more common mineral. Gypsum, goethite, malachite, chrysocolla, aurichalcite and azurite crust were not uncommon (Ransome, 1904a), though typically far more restricted in extent than calcite crusts as shown in Figure 2.7.9, where the malachite and azurite crust are modest in size compared to the calcite.

Drapery Calcite: Draperies occurred in many oxidation caves. As might be expected with the abundant copper and iron minerals present, the draperies were often very colorful as shown in Figure 2.7.10. It was common for individual draperies to contain white, green and red bands. Some draperies had serrated edges, caused by calcite crystal development along the bottom edge.

Flowers: Cave flowers were locally abundant in a number of areas in many of the oxidation caves. Gypsum was the only species recognized in this form and occurred as white, clusters or single ram's horn–shaped growths. An example is shown in Figure 2.7.11. Individual gypsum rams horns exceeded a meter in length, on rare occasion.

Flowstone: Flowstone was abundant and widely distributed in the oxidation caves. Several forms of flowstone were noted including monocrystalline, canopy shapes as well as the typical travertine form, as can be seen in Figure 2.7.12. The authors have seen banded travertine type flowstone more than three meters thick in several localities.

In the Bisbee oxidation caves, flowstone was composed of both calcite and aragonite but calcite flowstone was much more common. In some cases, the calcite was clearly a replacement of aragonite flowstone. Coloration by both copper and iron was commonplace for flowstone and often as alternating bands.

Frostwork: Aragonite commonly occurred as frostwork and was found in many oxidation caves, often covering large areas. Frostwork commonly developed as partial overgrowths on popcorn calcite (Figure 2.7.8 and 2.7.13).

Frostwork was usually colorless to white, but a few, bluish examples have been noted in a single occurrence in the Southwest Mine, as illustrated in Figure 1.5.8. A brown tinted example with included, finely divided goethite is shown in Figure 2.7.29.

Helictites: Both calcite and aragonite helictites are abundant in the oxidation caves at Bisbee, occurring as scattered clusters. Walls, ceilings and many of the speleothems were the base for strikingly beautiful displays of these erratic features (Figures 1.7.8, 1.7.21 and 2.7.14).

Like the other speleothems in Bisbee, helictites were often tinted by the ever-present copper minerals, further enhancing the already lovely displays with delicate shades of blue and green. Nearly the whole spectrum of helictic forms has been recognized. Vermiform and antler forms were the more common type while unicorn horn-like forms were the most unusual type, occurring in but a single locality (Figure 1.7.15).

Moonmilk: This nondescript speleothem is probably far more common than the single recognized occurrence would suggest. Nevertheless, moonmilk has been noted on a single specimen with aurichalcite. This specimen had been recovered from a cave in the Uncle Sam Mine more than 80 years ago.

Pearls: Cave pearls are surprisingly scarce at Bisbee. However, aragonite cave pearls have been recognized at a single locality in the Shattuck Mine. As a point of interest, post-mining cave pearl type deposits were quite common in abandoned mine workings throughout the whole mining area. Examples of exceptionally large, copper tinted, post-mining cave pearls are shown in Figure 2.7.15.

Rafts: Rafts were common in those few caves that contained pools. Figure 2.7.16 illustrates this nicely. Rafts must have also been abundant in many of the caves that no longer contain water as a few spots in a number of now-dry caves contain several centimeters of sunken raft litter. Many of the sunken rafts show subsequent, subaqueous calcite deposition as they are now several millimeters thick and have crystalline surfaces.

Most rafts were cream colored to white, but a few were tinted reddish to pink by iron. In Figure 2.7.17, pinkish raft litter is resting on similarly colored rhombohedral calcite cave spar.

Rims: Rims were found in a single cave in the Southwest Mine at Bisbee. This was confirmed by Hill (1981) in what is commonly called the "Higgins Mine Cave" (Hill & Forti, 1997). These are shown in Figure 2.7.18.

Rims require good airflow to form. Marginal cracks penetrating into with this cave, perhaps from the surface, must have provided the airflow since the cave is quite isolated from other known openings.

Shelfstone: Shelfstone formations were present in many caves, but usually restricted to a small area of the cave, typically in the lower part. Much of the shelfstone consisted of coarsely crystalline calcite spar with crystals up to 12 centimeters in length. See Figure 2.7.19 for such an example.

Much of the observed shelfstone was white, but a tan to yellowish to reddish coloration was seen in a number of places. No copper coloration of shelfstone calcite has been observed by the authors.

Tiered shelfstone (Figure 2.7.20) was common, reflecting the fluctuating water levels in the cave. As many as eight different shelfstone levels were found in one of the caves associated with the New Southwest Orebody.

Shield: At least two oxidation caves at Bisbee contained shield speleothems. In both cases, they were less than a meter across and lightly tinted green. One was in the cave described by Hovey (1911) and was named "The Elephant's Ear" by the miners.

Spar: Several varieties of spar have been recognized in oxidation caves. Pool spar was quite common as was subaerial spar. Many of the now dry pools in the caves contained calcite spar as rhombohedral, scalenohedral, nailhead or complex forms, often as large crystals. Examples of complex, tabular spar crystals are shown in Figure 2.7.21 while Figure 2.7.22 illustrates elongated rhombohedral crystals.

Spar frequently lined the whole pool area creating impressive crystal displays. Rhombohedral crystals were the most common with individual crystals up to 45 centimeters on an edge noted in one cave. Some stalactites that were partially submerged by rising water levels had massive spar deposits on their tips. Nailhead or scalenohedral spar masses formed "war club" occasionally exceeding 30 centimeters developed on the stalactite end. A small example is shown as item five in Figure 1.7.7. Pool spar was often colorless to white, but iron tinted spar, colored tan to brown and red-brown were wide spread as well.

Oddly, spar tinted by copper minerals was exceedingly rare in the caves, though it was occasionally found in the adjacent orebodies. That is unusual given the frequent occurrence of copper tinted stalactites, stalagmites, helictites and flowstone in many of the caves. Only two confirmed occurrences of spar tinted by copper minerals have been recognized and in both cases, the tinting was caused by tiny, included malachite crystals. In one of these occurrences, very lightly copper tinted spar was found as complex forms up to ten centimeters in length (Figure 2.7.21). The tinting was caused by tiny included, scattered malachite crystals.

Subaerial spar was a common, widely distributed form of calcite crystals that was found as a late stage, partial overgrowth on many speleothems. They were typically white to colorless crystals and less than a centimeter in length. Examples of this form of spar can be seen in Figure 2.7.23.

Selenite spar was found in a few caves usually as colorless, tabular crystals and less than two centimeters in length. However, in one instance, 15-centimeter selenite crystals were noted. In addition to being uncommonly large for Bisbee, these selenite crystals were also tinted light green by included malachite crystals.

Stalactites: The most abundant of speleothems in Bisbee's caves were stalactites and they seem to have occurred in almost every form. Some approached five meters in length.

Both aragonite and calcite stalactites were common with the calcite stalactites more abundant. Coloration by iron minerals was frequently observed, but it was the color often imparted by copper that made these forms so spectacular. While aragonite forms were often a delicate aqua blue, calcite could be found in hues of green from a light tint to a deep color.

Most of the colored stalactites were coarsely crystalline in structure, allowing a translucency that further enhanced the effect of the coloration.

Monocrystalline calcite stalactites, or calcite stalactites with crystalline terminations, were uncommon in Bisbee oxidation caves. A few of the monocrystalline stalactites developed as single, elongated crystals to 30 centimeters in length. An example is illustrated in Figure 1.7.7. Some stalactites with crystalline terminations were much longer as only the last five or so centimeters were in a crystalline form (See Figure 2.7.14). Crystal terminations were either a single crystal or a composite crystal made of a number of crystals in parallel or sub-parallel growth patterns.

Occasionally, goethite and malachite stalactites were found in these caves. Malachite stalactites in excess of a meter long occurred in several caves. In a very few instances, azurite stalactites were also found, but without exception these had formed as a result of azurite overgrowths on preexisting stalactites of malachite or goethite.

Stalagmites: Stalagmites were found in many of Bisbee's oxidation caves, but were far less abundant than stalactites. Indeed, many caves did not contain any stalagmites. This may be, in part, due to the floors of these caves often being unstable, shifting with the subsidence, as oxidation progressed or covered by falling breakdown material as the cave developed. Occasionally, the downward movement of the floor was uneven, causing some rotation of the stalagmites. Continued growth following movement resulted in a seemly bent stalagmite as shown in Figure 2.7.24.

Stalagmites too, were colored by the same elements that imparted hues to the stalactites, but were typically paler in color. Colorful stalagmites tinted by copper minerals were, in fact, relatively rare. One of the few examples known is the exceptional stalagmite on display in the mineral hall of the American Museum of Natural History in New York. It was once a part of the Bisbee oxidation cave display discussed in chapter four of part one and is illustrated in Figure 2.7.25.

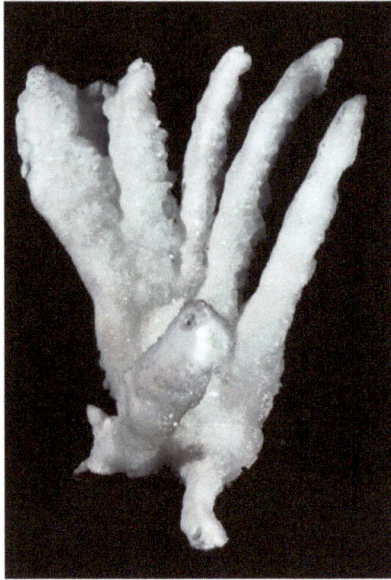

Figure 2.7.1: Aragonite anthodites with later calcite and minor malachite, 4th level, Southwest Mine, specimen - 13 centimeters.

Figure 2.7.2: Goethite boxwork in a cave wall, 4th level, Southwest Mine, vertical view - 1.1 meters.

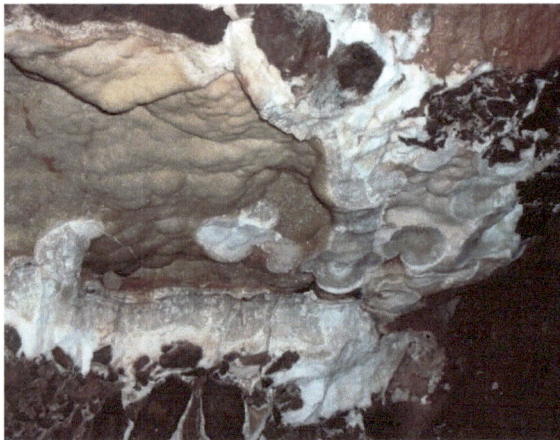

Figure 2.7.3: Calcite cave cloud formations 5th level, Southwest Mine, view - 1.5 meters.

Figure 2.7.4: Calcite cave cups to one cementer, 6th level, Southwest Mine, horizontal view - 15 centimeters.

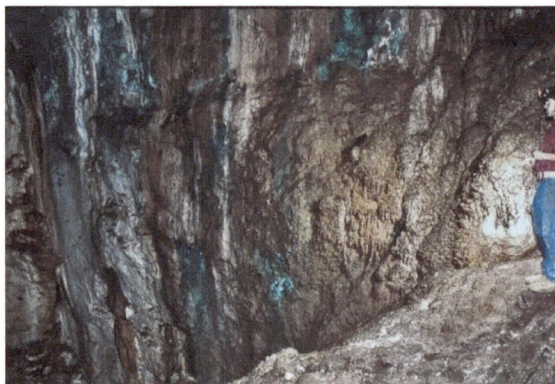

Figure 2.7.5: Coatings of malachite, chrysocolla and calcite on cave wall, 100 level, Holbrook Mine.

Figure 2.7.6: Azurite and malachite coatings on calcite stalactites, "A" level, Copper Queen Mine, view - 1.2 meters.

Figure 2.7.7: Two parallel lines of calcite columns of varying colors. These have formed under fractures in the cave ceiling that, have served as conduits for solutions, 7th level, Southwest Mine. Foreground horizontal view - five meters. Peter L. Kresan photograph taken in 1977.

Figure 2.7.8: Coralloids - calcite popcorn with aragonite frostwork, 7th level, Southwest Mine, view – 17 centimeters.

Figure 2.7.9: Crust of malachite with minor azurite to several centimeters thick, with calcite on a cave wall, 100 level, Holbrook Mine, vertical view - 3.2 meters.

150

Figure 2.7.10: Large, copper tinted calcite drapery forms on a cave wall, 100 level, Holbrook Mine, vertical view - 6.5 meters.

Figure 2.7.11: Gypsum cave flower, 6th level, Southwest Mine, view - nine centimeters.

Figure 2.7.13: Frostwork, aragonite on calcite, 4th level, Southwest Mine, view- 1.45 meters

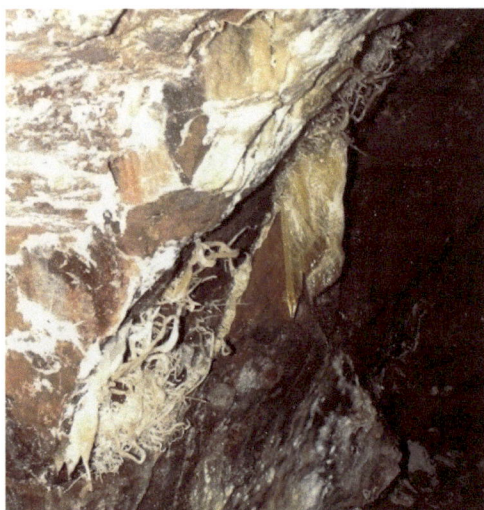

Figure 2.7.14: Nest of calcite helictites with a golden-colored, crystal terminated stalactite, above the New Southwest Orebody, 7th level, Southwest Mine, vertical view - 2. 4 meters.

Figure 2.7.12: Calcite flowstone engulfing stalagmites, 7th level, Southwest Mine, horizontal view - 3.2 meters.

Figure 2.6.15: Aragonite cave pearls of post-mining origin and tinted by copper. The largest is three centimeters in diameter, 2200 level, Campbell Mine.

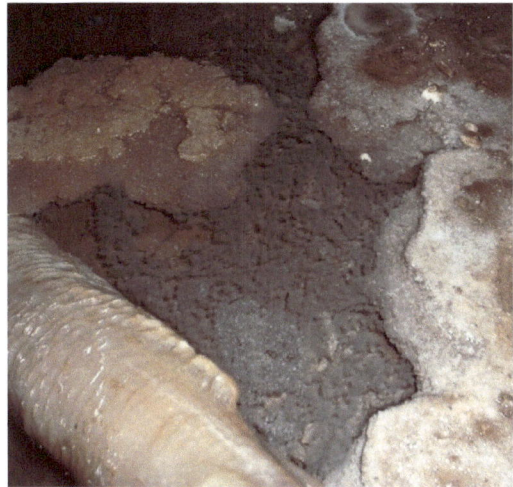

Figure 2.7.16: Calcite rafts floating on a small pool, 7th level, Southwest Mine, view - 85 centimeters.

Figure 2.7.17: Calcite raft litter on iron tinted rhombohedral "pool spar" calcite crystals, 6th level, Southwest Mine, view - 20 centimeters.

Figure 2.7.18: Several rim formations along a marginal crack, 7th level, Southwest Mine, view - 1.7 meters.

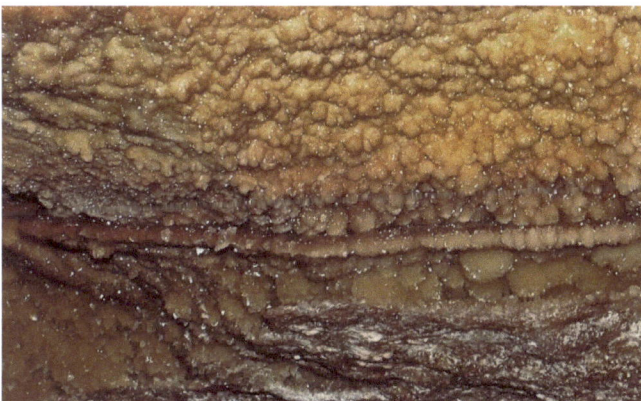

Figure 2.7.19: Left, linear shelfstone of coarsely crystalline calcite spar. Below the shelfstone, the area is covered by calcite crystals to three centimeters, 7th level, Southwest Mine, view - 1.8 meters.

152

Figure 2.7.20: Tiered, crescent shelfstone of iron tinted, coarsely crystalline, calcite spar reflecting two distinct solution levels, 6th level, Southwest Mine, horizontal dimension - 22 centimeters.

Figure 2.2.21: Colorless to white calcite spar as complex, tabular crystals on calcite tinted by included malachite, 6th level, Southwest Mine, view - 30 centimeters.

Figure 2.7.22: Calcite spar crystals to five centimeters, 5th level, Southwest Mine, view 11 centimeters. Bisbee Mining and Historical Museum specimen.

Figure 2.7.23: Calcite stalactites and subaerial spar forming, 7th level, Southwest Mine, horizontal dimension - ten centimeters. Peter L. Kresan photograph

Figure 2.7.24: Seemingly bent stalagmites in the 300 level Shattuck Cave. The bend is actually the result of the cave floor moving unevenly because of oxidation related subsidence. The slight rotation of the base with continued calcite deposition resulted in the bent appearance. Dimensions unknown, 1913.

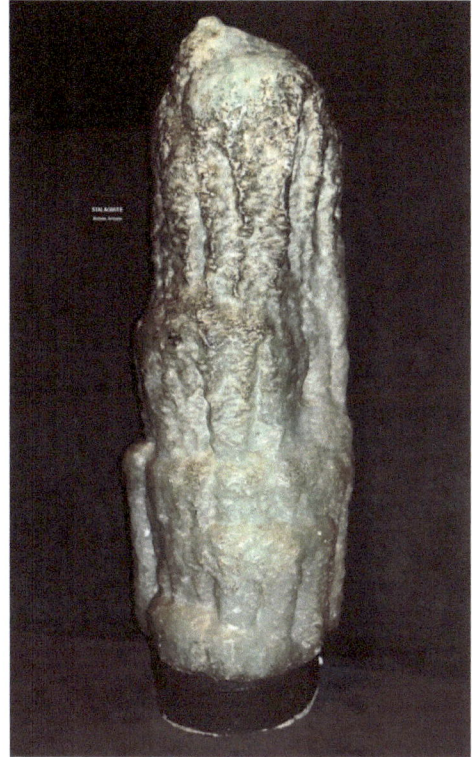

Figure 2.7.25: Copper tinted, calcite stalagmite 120 centimeters high, 4th level, Southwest Mine. This specimen is on display in the Mineral Hall of the American Museum of Natural History, New York. The dark color on the stalagmite is the result of being touched my many curious admirers over the years.

Speleothem Coloration: As has been noted above many times, it was common for speleothems in the oxidation caves at Bisbee to be colored, often intensely so. The stunning effect that copper frequently imparted to the coloration of these caves defies description. Much of the coloring by copper and iron was transported from mineralized areas above the cave with only minor amounts coming from the cave walls. Consequently, the caves that were in the highest positions with little, if any, mineralization above were typically less colorful than those lower in the system.

Copper colored both calcite and aragonite speleothems. This coloration was often two fold in origin: substitution of copper for calcium on a molecular level and/or simultaneous deposition of a copper mineral with the calcite or aragonite.

The substitution of copper for calcium within the calcite or aragonite crystal structure was the more common cause of coloration. In spite of this substitution of copper for calcium on the molecular level, the mineral remained calcite or aragonite and not a different minerals species. This is distinct from simultaneous deposition where two mineral species were co-deposited.

Substitution of calcium by copper produced a green tint in calcite and an aqua hue in aragonite. The degree of coloration in speleothem of either mineral was quite variable, ranging everywhere from just a slight tint to a medium green for calcite or aqua for aragonite. A good discussion of the substitution mechanism for coloration can be found in White (1981 and 1997).

The simultaneous deposition of a copper mineral occurring along with the deposition of calcite or aragonite, while far less common was not rare. In the instances where simultaneous deposition took place, the speleothem was usually composed of calcite and malachite was simultaneously deposited. Malachite deposition was not constant, but rather interspersed with periods of normal color calcite deposition.

Figure 2.7.26: Multi-colored calcite/aragonite stalactite section showing the green coloration of copper and red of iron in calcite as well as the blue-green color imparted by copper to aragonite, Czar Mine, specimen - 17 centimeters. USNMNH specimen.

Figure 2.7.27: Polished section of a calcite stalactite, showing calcite co-deposited with malachite as flecks to 0.5 millimeter imparting a strong green tint, Czar Mine, view – eight centimeters.

Malachite inclusions have been observed in three forms: as scattered, minute, acicular crystals or as small, thin, flake-like particles or patchy crust of less than a millimeter in thickness. The intensity of coloration of the speleothem was determined by the density of malachite inclusions. Calcite speleothems exhibited every shade of green from a delicate tint to dark green. Sections were so dark that it was difficult to determine, at a glance, if the specimen was malachite or deeply colored calcite.

Interestingly, calcite speleothems colored by included azurite have not been seen by the authors. To be sure, the occasional calcite speleothem may have a few small azurite crystals on the surface, but never included to any degree.

The aqua blue coloration of aragonite speleothems is invariably a result of copper substitution of calcium in the aragonite crystal structure. The intensity of coloration was quite variable with light blue-green tints far more common than the deep aqua color. In a few rare instances rosasite or small amounts of azurite and in one case, conichalcite were simultaneously deposited as the aragonite speleothems formed. Both rosasite and azurite gave a bluish color to aragonite, while conichalcite imparted a yellowish green hue.

Figure 2.7.28: Calcite stalactite showing coloration by copper and iron caused by co-deposition. In the case of iron, goethite was deposited as micro flecks on the surfaces as the stalactite grew and were included in the calcite. Copper coloration is the result of tiny particles of malachite deposited simultaneously with the calcite. As can be seen, there were periods of calcite deposition without any coloration, 4th level, Southwest Mine, view - 8.2 centimeters.

Both calcite and aragonite speleothems were colored by including fine particles of iron oxides. However, it was far more common to find calcite tinted by iron minerals than aragonite. Yellow-browns to red-browns were common hues in many of the speleothems, particularly flowstone. Variable degrees of coloration by iron minerals in the same speleothem were common.

Figure 2.7.29: Aragonite on calcite. The aragonite crystals are partially tinted by included goethite, 3rd level, Southwest Mine, view - nine centimeters.

In spite of the abundance of manganese oxides in the Bisbee oxidation caves, these minerals colored few speleothems. In two locations, flowstone was seen that had been locally colored a dark chocolate brown by the inclusion of manganese oxides.

There can be little doubt that some of the coloration in many of the tan to brown speleothems was due to the inclusion of organic (humic) substances. However, because of the overwhelming presence of iron, the effects of this agent were almost always masked. Figure 2.6.29 illustrates what is believed to be color imparted by humic substances to speleothems.

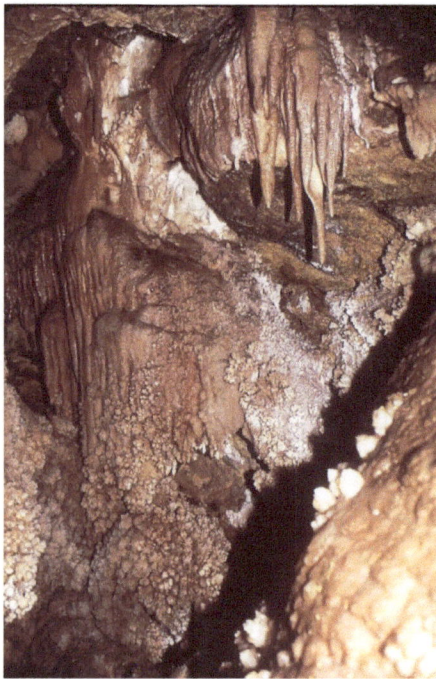

Figure 2.7.29: Calcite speleothems possibly colored by humic substances, as the cave is quite close to the surface, 4th level, Southwest Mine, vertical view - 1.3 meters.

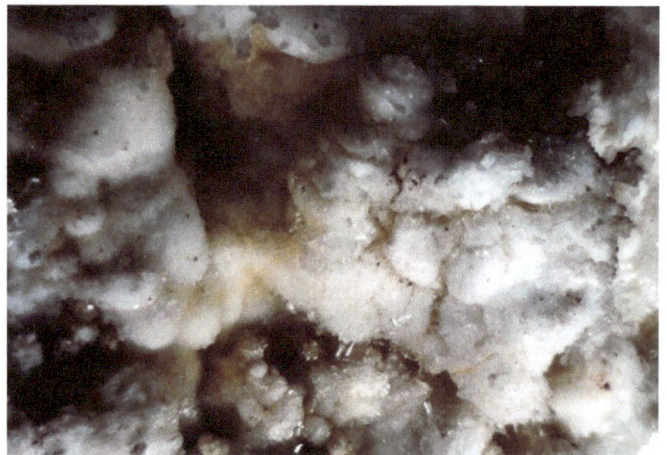

Figure 2.7.30: Minor, colorless gypsum as two-millimeter crystals on calcite under normal light, 300 level, Shattuck Mine, view - four centimeters.

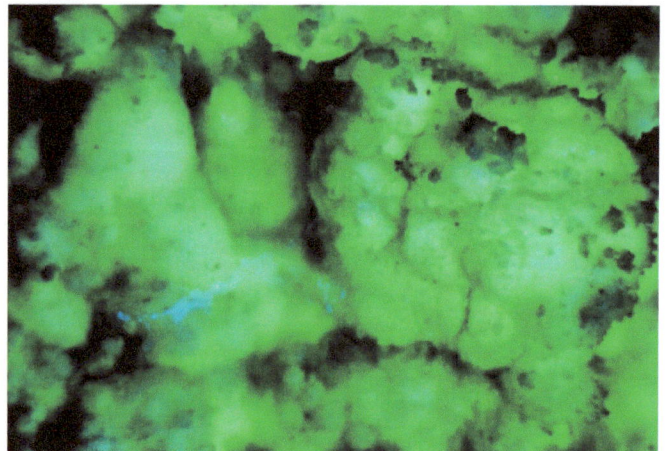

Figure 2.7.31: The same area as shown above under shortwave ultraviolet light. The gypsum is not responding as only the underlying calcite is fluorescing.

Fluorescence in speleothems: A number of oxidation caves contained speleothems that exhibit some degree of fluorescence. However, fluorescence is far from ubiquitous and partial fluorescence in a speleothem is a common feature. A green fluorescence was the more typical and the green color was often quite vivid. Few speleothems exhibited white to blue-white fluorescence and in one case, an orange response was noted by the authors in pieces of flowstone collected from the waste dump of the 6th level Southwest Mine. No speleothems fluoresced orange-red to red, which is not too surprising as this is something more typical of hydrothermal deposition, not normal cave growth.

Chapter Eight: Minerals Recognized in the Oxidation Caves at Bisbee

A number of mineral species have been recognized in the oxidation caves at Bisbee. As would be expected, they are all supergene in origin. Identification of the minerals listed below has been through the relevant literature, visual identification or physical test of the common species and x-ray diffraction for the rare and unusual species as denoted by an asterisk. The species are:

aragonite	chrysocolla	gypsum	murdochite*	smithsonite
aurichalcite	conichalcite	hematite	plattnerite*	stolzite*
azurite	cuprite	hemimorphite	quartz	volborthite*
bromargyrite*	descloizite*	hetaerolite	rosasite	
calcite	dolomite	malachite	romanèchite	
cerussite	goethite	mimetite	sepiolite*	
chalcophanite	gold	mottramite*	siderite	

Table 2.8.1: A list of minerals recognized in the oxidation caves at Bisbee.

The occurrences of these minerals in the caves at Bisbee are described below. Additional information on these and other minerals at Bisbee can be found in Graeme (1981 and 1993). Formulas for the minerals are from Fleischer and Mandarino (1995). Following the description of the minerals are photographs of those species that have not previously been illustrated or of examples which manifest other, notable characteristics.

Aragonite $CaCO_3$: Aragonite was the second most abundant mineral found in the oxidation caves following calcite. Many of the speleothems in the caves at Bisbee were composed of aragonite with stalactitic forms far and away the more typical occurrence, as stalactites were the more common speleothem in these caves. Aragonite stalactites and stalagmites were often in excess of several meters in length or height and frequently tinted by copper or iron minerals.

Microcrystalline to coarsely crystalline aragonite flowstone occurred as compact, thick, reniform coverings of the cave floor and walls. Aragonite flowstone was often colored by copper and iron and exhibited color bandings with occasional white or colorless zones. The banding reflects the varying phases of the aragonite growth and coloration that occurred in the cave, reflecting the changing nature of the depositional solutions. Interlayered bands of calcite commonly occurred and were probably formed by alteration as calcite readily replaces aragonite, with the surviving aragonite possibly preserved by metal ions.

Coralloidal growths of aragonite, termed flos-ferri, occurred as scattered patches, generally in the upper portions of the openings and frequently associated with stalactites of both aragonite and calcite. These formations were a later stage of deposition that developed on earlier aragonite speleothems. Here too, coloration by copper was common. An example is shown in Figure 1.7.7. Acicular aragonite (frostwork) frequently occurred in many of the oxidation caves and openings. The acicular aragonite often exhibited oriented growths on botryoidal calcite.

The aragonite crystals, while typically small and acicular, could be quite large. Chisel-shaped crystals of five centimeters on large botryoidal calcite were occasionally found. Stalactitic like, jackstraw masses of three to five centimeters chisel shaped crystals in excess of a meter long were found in one cave. Acicular and chisel shaped crystals of aragonite are almost always colorless to white with a small number of iron or copper tinted specimens known as illustrated in Figures1.5.8 and 2.7.29, respectively.

The rarest form of aragonite in the oxidation caves at Bisbee was as helical growths. Some of these helical growths were reminiscent of a machine screw (see figure 1.7.16). These most unusual speleothems were found as randomly oriented protrusions from the walls and ceilings of several small oxidation caves. With a maximum length of 18 centimeters, these aragonite forms are typically restricted to one small area of the cave and occur as individual growths or clusters, often adjacent to more typical aragonite formations.

Calcite paramorphs after aragonite have been noted throughout the oxidation caves in Bisbee by the authors.

Aurichalcite $(Zn,Cu)_5(CO_3)_2(OH)_6$: Far and away, the most common mode of occurrence for this lovely species at Bisbee was in oxidation caves. Aurichalcite occurred as a partial covering of the walls or as a lining of voids between boulders in the lower parts in a small number of these caves. The bottoms of the oxidation caves were almost always a chaotic mass of boulders of iron oxides and/or limestone. In the caves that contained aurichalcite, the voids between the boulders were often lined with this mineral. This type of occurrence was probably noted by Kunz (1885) when he described it as "in beautiful tubes lining cavities."

Azurite $Cu_3^{+2}(CO_3)_2(OH)_2$: The strikingly beautiful mineral, azurite, was found in a number of oxidation caves, though rarely in large amounts, with the extraordinary occurrence in the Southwest Orebody, the lone exception . Azurite occurred as localized, thin crust on cave walls in the lower portions of the caves and as crystals on malachite (Ransome, 1904a). In one small cave, it was the dominant species with the walls and floors covered by crystalline azurite and minor malachite and cuprite (Graeme, 1981). On rare occasion, azurite was found in very minor amounts as scattered, very small crystals on calcite speleothems.

Bromargyrite AgBr: Bromargyrite was an important economic mineral in the "Lead Cave" Orebody in the Shattuck Mine. An example is illustrated in Figure 2.8.1. It was found as waxy, greenish-yellow, patchy crust on rock fragments with the cerussite ores. At the edge of the cave, it was found as partial to complete replacements of silver, including distinct pseudomorphs, on siliceous fragments and as one millimeter, well-formed modified cubic crystals. Other minerals found with bromargyrite in the "Lead Cave" were tiny crystals of specular hematite, as well as plattnerite and murdochite.

Calcite $CaCO_3$: Calcite was the most abundant mineral in the vast majority of the oxidation caves at Bisbee. Calcite speleothems were everywhere in these caves, often as spectacular displays. Many caves contained calcite in crystal form. A good deal of the calcite found in the oxidation caves at Bisbee was actually a replacement of aragonite, a common and well-documented occurrence worldwide.

Cerussite $PbCO_3$: The "Lead Cave," in the Shattuck Mine contained several tens of thousands of tons of cerussite as spongy to sandy material of a, tan to reddish color and mixed with minor amounts of specular hematite. Locally, cerussite occurred as loosely cemented in porous masses of tan to brown intergrown crystals from two to six millimeters. Also, cerussite occurred as boulders of fine-grained material with cores of anglesite, grading in to unaltered galena. Typical material from the "Lead Cave" is illustrated in Figure 2.8.2. This same area produced a few groups of two centimeter, sixling-twinned crystals of cerussite from near the limestone contact that were coated by an unidentified, black manganese oxide.

Chalcophanite $(Zn,Fe^{2+},Mn^{2+})Mn_3^{4+}O_7 \cdot 3H_2O$: A small cave in the top of a crosscut on the 1400 level of the Cole contained large amounts of massive, botryoidal and stalactitic chalcophanite. It was often covered by bright drusy overgrowths of the same mineral. Boxwork goethite was also encrusted by drusy chalcophanite at this locality. See Figure 2.8.3.

Chrysocolla $(Cu^{+2},Al)_2H_2Si_2O_5(OH)_4 \cdot nH_2O$: Chrysocolla was abundant in the oxide copper deposits found in the cave bottoms and less commonly as a localized crust in the upper parts. Chrysocolla usually formed as a replacement of earlier copper minerals, particularly cuprite and malachite. A fine example is shown in Figure 2.8.4.

Conichalcite $CaCu^{2+}(AsO_4)(OH)$: Spotty crust of conichalcite with small calcite crystals occurred along a cave wall on the 7th level of the Southwest Mine. In the same cave, tiny, scattered yellow-green conichalcite crystals on altered limestone were noted by the authors. A minor amount of mimetite was associated with conichalcite at this locality. Conichalcite imparted a light, yellow-green hue to a small amount of the calcite in this cave.

Cuprite Cu_2O: Many of the oxidation caves at Bisbee contained massive cuprite, in small amounts and in the oxide mass in the cave walls and bottoms (Douglas, 1881), usually with a thick alteration rind of chrysocolla and malachite. In a few rare instances, it was occurred as tiny (one millimeter – two millimeters) crystals on other copper minerals, notably azurite on cave walls (Bates, 1896), (Graeme, 1983). One small cave in the Irish Mag Mine contained large amounts of cuprite as drusy crust on stubby goethite stalactites and as coatings on the walls.

Descloizite $PbZn(VO_4)(OH)$: Descloizite has been recognized in the "Shattuck Vanadium Cave" above the 600 level of the Shattuck Mine as scattered, tabular, minute crystals on siliceous fragments from the cave bottom on mottramite and with later calcite. In the Southwest Mine, modest amounts of tiny descloizite crystals occurred in voids between boulders in the bottom of a large cave and partially overgrown by mimetite with malachite and minor plattnerite. In the Higgins Mine, a very few examples of crystalline black/brown descloizite with calcite were found in a small cave. An example is shown in Figure 2.8.5

Dolomite $CaMg(CO_3)_2$: The most notable occurrence of dolomite in a Bisbee oxidation cave was as partial, white overgrowths on acicular aragonite in a large cave on the 7th level of the Southwest Mine. For some unexplained reason this white dolomite that became tan in color after several years of exposure to the surface environment. An example is shown in Figure 2.8.6.

Goethite $Fe^{+3}O(OH)$: Goethite was the most abundant non-carbonate mineral in the oxidation caves at Bisbee. Goethite occurred as the principal constituent of the sulfide oxidation products throughout the cave. The cave's walls and floors as well as the oxide masses underneath these caves held huge amounts of this material as an impure mixture with hematite and other supergene derived minerals.

It was common for goethite to form a large part of the walls of the caves. However, it was typically coated by calcite or aragonite overgrowths and therefore not obvious. Stalactitic and botryoidal goethite was found in many caves, but theses too were frequently calcite coated. Boxwork goethite was abundant in the Bisbee caves.

Gold Au: A unique occurrence for gold was in a cave adjacent to a silica breccia in the Uncle Sam Mine. It was found as small flakes with rounded edges and tiny nugget-like forms in loose, lightly iron stained quartz sand in the cave bottom. This gold was well rounded with the appearance of having been water worked (Bateman & Murdoch, 1914). There is little doubt that this gold was present either in the primary ores or the silica breccia prior to the formation of the cave.

Gypsum $CaSO_4 \cdot 2H_2O$: Gypsum was found in a great many of the caves encountered during mining. In these openings, the more typical form was "ram's horn" growths. While they typically extruded from the walls or ceilings of the caverns, a few have been found growing on the floor. The gypsum formations were generally, less than ten centimeters in length, some were found that approached a meter in length. A 40-centimeter specimen was salvaged from a cave in the Holbrook Mine in 1904 and is shown in Figure 2.8.7. On rare occasion; copper would impart a pale blue-green to green hue to these formations. Small, colorless gypsum spar was noted in some of the caves by the authors, with these crystals usually found in the lowest parts of the cave.

Also, found in a few cave bottoms were blocks of very porous, almost sponge-like, intergrowths of small gypsum crystals. These blocks were up to 1/2 meter high and slightly less wide. They were both attached to the cave floor and free standing. Often minor amounts of malachite were randomly included in the gypsum blocks. Vivid green fluorescence was a common feature of many of these gypsum blocks.

Hematite Fe_2O_3: Hematite was a common constituent of the oxide masses associated with the oxidation caves. It was invariably mixed with the more abundant goethite and was typically a soft, ocherous material. Much less common, hematite occurred in the cave walls with goethite. Hard, compact hematite was frequently associated with the oxide ores well below the cave bottom.

Hemimorphite $Zn_4Si_2O_7(OH)_2H_2O$: A number of oxidation caves contained modest amounts of hemimorphite as tiny radiating clusters of minute, colorless crystals. It typically occurred with other zinc minerals, particularly rosasite.

Malachite $Cu_2^{+2}(CO_3)(OH)_2$: Malachite was the most common copper mineral in oxidation caves at Bisbee. It often occurred as beautiful crust of fibrous crystals (Ransome, 1904a). Botryoidal malachite, often occurring with azurite, was common in the cave bottoms and also as a lining of boxwork vugs and voids between boulders. On rare occasion, stalactites up to a meter in length were found.

The oxide masses below the caves often contained large boulders of spongy malachite and immense quantities of fibrous malachite as veinlets and void linings. The economic importance of these types of deposits to Bisbee's success cannot be overstated.

Mimetite $Pb_5(AsO_4)_3Cl$: Small amount of tiny yellow mimetite crystals with partial overgrowth of descloizite and with calcite were recognized in a cave bottom in the Southwest Mine.

Mottramite $PbCu^{2+}(VO_4)(OH)$: The "Shattuck Vanadium Cave" above the 600 level of the Shattuck Mine was decorated largely with velvet-like to drusy mottramite. It occurred as brown-to-brown-black crust, popcorn-like growths and stalactites up to several centimeters in length

(Wells, 1913). Though Wells (1913) reported it as cuprodescloizite (descloizite) from this locality, it was subsequently shown to be mottramite (Taber & Schaller, 1930).

Murdochite $PbCu_6^{+2}O_{8-x}(Cl,Br)_{2x}$: Murdochite was noted by the authors as scattered, 0.5 millimeter, splendent, black crystals on rock fragments in a cave bottom on the 300 level of the Shattuck Mine. It was also found in minor amounts in a cave on the 100 level of the Holbrook Mine associated with plattnerite.

Plattnerite PbO_2: Plattnerite has been recognized in several, widely separated caves, always in small amounts and as tiny, submetallic, black crystals (see Figure 2.8.8).

Quartz SiO_2: Small amounts of quartz, both crystalline and cryptocrystalline, have been found in a number of caves. It is far more common in the caves associated with the siliceous breccias of the Southwest and Shattuck mines.

Rosasite $(Cu^{+2}, Zn)_2(CO_3)(OH)_2$: This lovely light blue mineral has been recognized in several caves in small amounts as tiny spheroids on goethite.

In one unusual instance in the Uncle Sam Mine, rosasite was found as velvet, botryoidal linings up to a centimeter thick in pockets in the walls and in boulders in a small oxidation cave. Aurichalcite occurred in these pockets as a partial overgrowth on the rosasite. An example can be seen in Figure 2.8.9.

Romanèchite $(Ba,H_2O)(Mn^{4+},Mn^{3+})_5O_{10}$: Much of the soft black mud-like material in many of the oxidation caves of the Southwest, Czar and Holbrook mines was romanèchite.

Sepiolite $Mg_4Si_6O_{15}(OH)_2 \cdot 6H_2O$: Numerous small, rounded, spongy pieces and compact nodules of sepiolite occurred in the bottom of a small oxidation cave on the 6th level of the Southwest Mine. In places, this sepiolite was overgrown or connected into clusters by a thin crust of dolomite. A light, green fluoresces is characteristic of these specimens.

On the 7th level of the Southwest Mine, one part of a very large oxidation cave contained many tons of loose, very light almost popcorn-like pieces of sepiolite. The sepiolite filled the bottom of the cave to several meters. Acicular aragonite occurred as overgrowths on a small amount of this material.

Siderite $FeCO_3$: Siderite was abundant in the early stages of cave formation, but relatively uncommon in fully developed oxidation caves, as continued oxidation converted it to goethite (Trischka, et al., 1929). When found, it comprised large masses of boxwork material of iridescent, botryoidal and stalactitic forms. Typically, drusy, dark brown siderite occurred as overgrowths on compact tan siderite.

Smithsonite $ZnCO_3$: Smithsonite was occasionally a constituent of the oxides underlying some of the caves at Bisbee and much less commonly, it occurred as spotty crust on the cave walls. It was usually found as yellow to white to gray boxwork forms in the oxide masses and as porous, yellowish crust in the caves.

Stolzite $PbWO4$: Specimens of cerussite on romanèchite from a small cave on the 6th level of the Southwest Mine contained a few 0.1 millimeter, red-orange, dipyramidal crystals of stolzite with very minor hemimorphite and plattnerite. An example is illustrated in Figure 2.8.10

Volborthite $Cu_3^{2+}V_2^{5+}O_7(OH) \cdot 2H_2O$: Volborthite occurred in the "Shattuck Vanadium Cave" on the 9th floor, or at 78 feet above the 600 level. It was found here in small amounts as tiny, amber to yellowish, pseudohexagonal crystals with mottramite and minor calcite.

Figure 2.8.1 Bromargyrite with calcite and minor malachite from the "Lead Cave," 300 level, Shattuck Mine, view - two centimeters.

Figure 2.8.2: Tan and gray, granular cerussite from the "Lead Cave," 300 level, Shattuck Mine, view - three centimeters.

Figure 2.8.3: Chalcophanite, 1400 level, Cole Mine, view - seven centimeters.

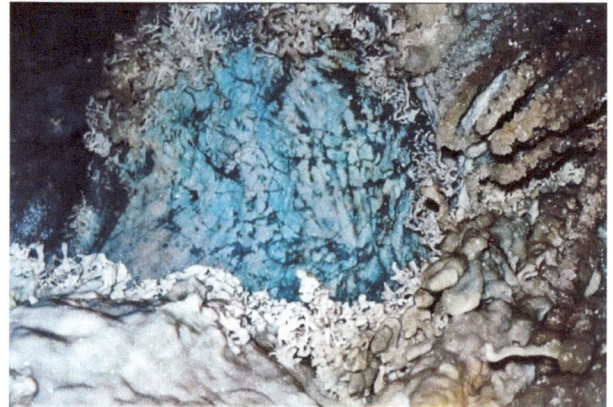

Figure 2.8.4: A meter-wide mass of blue-green chrysocolla with patches of black chrysocolla, surrounded by helictic calcite on a cave wall, 100 level, Holbrook Mine.

Figure 2.8.6: Descloizite, Tunnel level, Higgins Mine, view - 1.5 centimeters.

Figure 2.8.5: Cuprite crystals to four millimeters on stalactitic goethite, Irish Mag Mine.

Figure 2.8.7: Tan dolomite overgrowths on aragonite, 7th level, Southwest Mine, vertical view - six centimeters. Peter L. Kresan photograph.

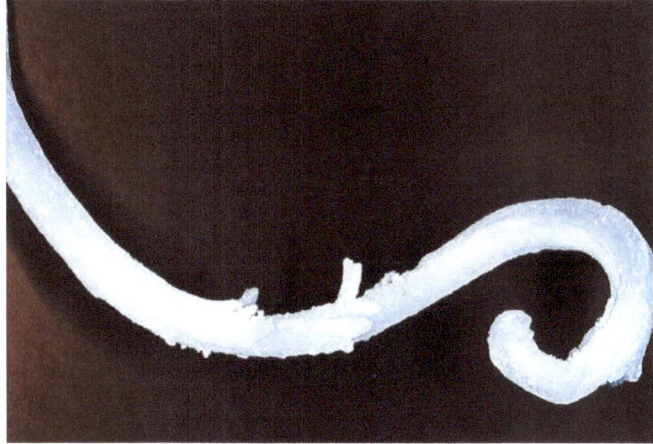

Figure 2.8.8: Gypsum, variety selenite, rams-horn growth, Holbrook Mine, specimen - 40 centimeters. Bisbee Mining and Historical Museum collection.

Figure 2.8.9: Plattnerite crystals to two millimeters on altered limestone, 7th level, Southwest Mine.

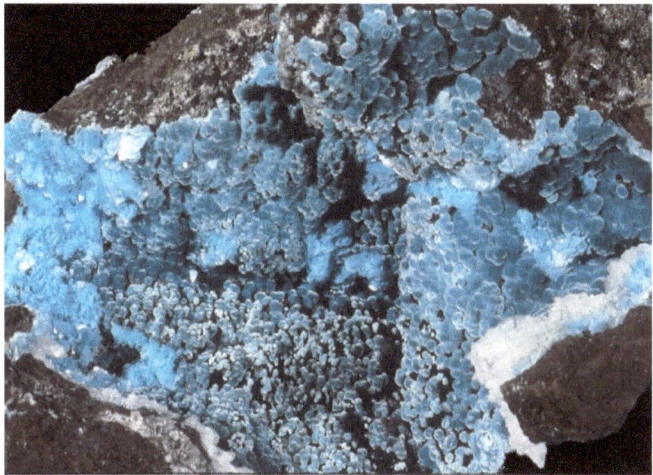

Figure 2.8.10: Botryoidal rosasite and acicular aurichalcite with calcite lining a 20 centimeter pocket in a boulder of mixed manganese oxides, "B" level, Uncle Sam Mine.

Figure 2.8.11: Orange stolzite and white cerussite on romanèchite, 6th level, Southwest Mine, view - 1.2 centimeters.

166

BIBLIOGRAPHY

AMERICAN JOURNAL OF SCIENCE, THE, (1891) Advertisement by Geo. L. English & Co. **42,** No.247-252.

AMERICAN JOURNAL OF SCIENCE, THE, (1917) Advertisement by Albert H. Petereit. **44,** No. 262.

AMERICAN MUSEUM JOURNAL, THE, (1914) *Museum notes.* **14,** p 216.

AMERICAN MUSEUM OF NATURAL HISTORY, THE, (1918) Annual Report for the Year of 1917. p. 187.

AMERICAN MUSEUM OF NATURAL HISTORY, THE, (1911) Forty Second Annual Report for the Year of 1910, p. 46.

AN OCCASIONAL CORRESPONDENT, (1887) Earthquake Phenomena in Arizona, *The Engineering and Mining Journal,* **47,** p 417–418.

ARIZONA DAILY STAR, 1897 *A Masonic Souvenir,* December 25, **39,** p. 293

ARIZONA REPUBLICAN, THE, (1897) *The Cave at Bisbee,* November 3, p.1.

ARIZONA REPUBLICAN, THE, (1897) *Bisbee's Big Week,* November 14, **8,** No. 145, p. 1.

ARIZONA REPUBLICAN, THE, (1898) *All over Arizona,* February 13, p. 3.

ARIZONA SILVER BELT, (1916) *Delegates to Elk's Reunion,* March 25, p.2.

AULER A. S. and SMART P. L., 2004. Rates of condensation corrosion in speleothems of semi-arid northeastern Brazil. *Speleogenesis and Evolution of Karst Aquifers,* 2 (2) December 2004, 2

BANCROFT, H. E., (1893a) *The Book of the Fair, an Historical and Descriptive Presentation of the World's Science, Art, and Industry, as viewed through the Columbian Exposition at Chicago in 1893.* The Bancroft Company, Publishers, Chicago and San Francisco, 487 p.

BANCROFT, H. E., (1893b) *The Book of the Fair, an Historical and Descriptive Presentation of the World's Science, Art, and Industry, as viewed through the Columbian Exposition at Chicago in 1893.* The Bancroft Company, Publishers, Chicago and San Francisco, 832 p.

BATEMAN, M. N. and MURDOCH, J., (1914) Secondary enrichment investigations; notes on Bisbee, Arizona. Unpublished notes, Harvard University Mineralogical Museum files, 240 p.

BATES, A. C., (1896) News and comments. *The Mineral Collector,* **2,** No. 12.

BATES, A. C., (1902) Minerals. *Popular Science News,* **34,** No. 4, 85.

BEASLEY, W. L., (1916) Copper Queen cave in New York. *The Engineering and Mining Journal,* **102,** 379-380.

BISBEE DAILY REVIEW, (1904) *World's Fair Edition.* Sam'l F. Myerson Printing Co., St Louis 96 p.

BISBEE DAILY REVIEW, (1906) *Find of fossils in the Holbrook Mine.* January 13.

BONILLAS, Y. S., TENNEY, J. B., and FEUCHERE, L., (1916) Geology of the Warren mining district, *A.I.M.E. Transactions,* **55,** 285-355.

BOTTRELL, S. R., GUNN, J., and LOWE, D. J., (2000) Calcite Dissolution by Sulfuric Acid: in *Speleogenesis, evolution of karst aquifers* edited by A. B. Klimchouk, D. C. Ford, A. N. Palmer and W. Dreybrodt. National Speleological Society, Inc., Huntsville, Alabama, 156 p.

BOSÁK, P., (1989) An introduction to karst-related mineral deposits; in *Paleokarst, A Systematic and Regional Review.* P. Bosák, D. C. Ford, J. Glazek and I. Horáček editors. Eselvier & Academia, Amsterdam and Praha, 367-375.

BRINSMADE, R. B., (1907) Copper mining at Bisbee, Arizona. *Mines and Minerals,* **27,** 289-293.

BROOK, G. A., BURNEY, D. A. & COWERT, J. B., (1990) Desert paleoenvironmental data from cave speleothems with examples from the Chihuahuan, Somali-Chalbi and Kalahari deserts. Paleogeography, Paleoclimatology, Paleoecology, **76.**

BRYANT, D. G., and METZ, H. E., (1966) Geology and ore deposits of the Warren mining district; in *Geology of the Porphyry Copper Deposits, Southwestern North America,* edited by S. Titley and C. Hicks. The University of Arizona Press, Tucson, Arizona, 189-204.

CALUMET and ARIZONA MINING COMPANY (1910) Annual report for the year of 1909.

CHICAGO DAILY (1882) *Mining Exposition.* August 17, 6.

CHICAGO DAILY TRIBUNE, (1891) *Exposition Notes.* March 13, p.6.

CHICAGO DAILY TRIBUNE, (1893) *Wealth of Arizona Mountains.* April 24, p. 3.

COLE, M. E., (1908) *Jottings from Overland Trip to Arizona and California.* Poughkeepsie, New York: By the author, 99 p.

CORONET MEMORIES, (1899) *Log of Schooner-Yacht Coronet off shore cruses from1893 to1899.* F. Tennyson Neely, Publisher, London, New York.

CRAFT, M. C., (1899) *The Strangest Masonic Lodge-room in the World.* Leslies Weekly, **88**, No.2280.

CURTIS, J. S., (1884) Silver-lead deposits of Eureka Nevada. *U. S. Geological Survey Monographs, volume VII.* Government Printing Office, Washington, D.C. 200p.

DeWILDE, E. J., (1915) Brief notes on copper deposits of Bisbee, Arizona. *Mining and Engineering World,* **42,** 583-585.

DOUGLAS, J., (1881) Report on the Copper Queen Mine. Letter to Prof. Silliman dated February 8, 1881, 7 p. Arizona Historical Society files.

DOUGLAS, J., (1900) The Copper Queen Mine, Arizona. *A.I.M.E. Transactions,* **29,** 511-546.

DUBLYANSKY, V. N. and DUBLYANSKY, Y. V., (2000) The role of condensation in karst hydrogeology and speleogenesis. In: Klimchouk A. B., Ford D.C., Palmer A. N., Dreybrodt W. (eds.): *Speleogenesis: Evolution of Karst Aquifers.* National Speleological Society, pp. 100-112.

DUBOIS, S. M., and SBAR M., (1981) The 1887 Earthquake in Sonora: Analysis of regional ground shaking and ground failure; in *Proceedings of Conference XIII, Evaluation of Regional Seismic Hazards and Risk*, edited by W. W. Hays and compiled by B. B. Charonnat, *U.S. Geological Survey. Open File Report. 81-437*, Reston, VA, 1981.

DŹULYŃSKI, S. & SASS-GUSTKIEWIEZ, M., (1989) Pb-Zn ores; in *Paleokarst, A Systematic and Regional Review.* P. Bosák, D. C. Ford, J. Glazek and I. Horáček editors. Eselvier & Academia, Amsterdam and Praha, 377-397.

DEPARTMENT OF THE INTERIOR, (1898) Annual reports for the year ending June 30, 1898. Government Printing Office, Washington, D. C., 271-273.

EMMONS, S. F., (1886) Geology and Mining Industry of Leadville, Colorado with Atlas. *Monographs of the United States Geological Survey, Volume XII.* Government Printing Office, Washington, D.C., 770 pages and 45 plates.

EMMONS, W. H., (1913) The Enrichment of Sulfide Ores. *U.S. Geological Survey Bulletin* **529,** Government Printing Office, Washington, D.C., 260 p.

EMMONS, S. F., IRVING, J. D., and LOUGHLIN, G. S., 1927, Geology and Ore Deposits of the Leadville Mining District, Colorado: *U.S. Geological Survey Professional Paper* **148,** Government Printing Office, Washington, D.C., 378 p.

ENGINEERING AND MINING JOURNAL, THE, (1881) Cave struck. *The Engineering and Mining Journal,* **31,** 215.

ENGINEERING AND MINING JOURNAL, THE (1884) Unique specimen of silver ore. *The Engineering and Mining Journal,* **38,** 270.

ENGINEERING AND MINING JOURNAL, THE, (1883) Cave with large volumes of copper ore is found. *The Engineering and Mining Journal,* **36,** 69.

ENGINEERING AND MINING JOURNAL, THE (1893) The Arizona Exhibit. *The Engineering and Mining Journal,* **55,** 533.

ENGINEERING AND MINING JOURNAL, THE, (1910a) The Mining News - Arizona, Copper Queen. *The Engineering and Mining Journal,* **89,** 436.

ENGINEERING AND MINING JOURNAL, THE (1910b) New ore found in the Calumet & Arizona mines. *The Engineering and Mining Journal,* **89,** 935.

ENGINEERING AND MINING JOURNAL, THE, (1917) The Mining News - Arizona, Shattuck and Arizona. *The Engineering and Mining Journal,* **103,** 688.

ENGLISH, G. L., (1890) *Catalogue of Minerals for Sale by Geo. L. English & Co.,* Geo. L. English Co., New York and Philadelphia, 100 p.

FLEISCHER, M. and MANDARINO, J. A., (1995) *Glossary of Mineral Species 1995.* The Mineralogical Record Inc., Tucson, Arizona, 267 p.

FORD, D. C., and HILL, C. A., (1989) Dating results from Carlsbad Cavern and other caves in the Guadalupe Mountains, New Mexico. *Isochron/West,* No. 54, 3-7.

FORD, D. C., and HILL, C. A., (1999) Dating of speleothems in Kartchner Caverns, Arizona. *Journal of Cave and Karst Studies,* 61 (2): 84-88.

GRAEME, R. W., (1981) Famous mineral localities: Bisbee, Arizona. *Mineralogical Record,* **12,** 258-319.

GRAEME, R. W., (1993) Bisbee revisited, an update on the mineralogy of this famous locality. *Mineralogical Record,* **24,** 421-436.

GRAEME, R. W., Graeme, D. L. Graeme, R. W. IV, (2015) An update on the Minerals of Bisbee, Cochise County, Arizona. *Mineralogical Record,* **46,** 627-641.

GRAND LODGE OF ARIZONA, (1897) *Proceedings of the M. ∴ W. ∴ Grand Lodge of Free and Accepted Masons of the Territory of Arizona, at its Sixteenth Annual Communication held at Masonic Hall, in the Town of Bisbee And of a Session held in the*

Cave of the Copper Queen Mine, commenced on Tuesday, November 9th, A. D. 1897, A. L. 5897, and terminated on Thursday, November 11, A. D. 1897, A. L. 5897. Arizona, 1897, 35-36.

HARVARD COLLEGE, (1915) Annual Report of the Director of the Museum of Comparative Zoology at Harvard College to the President and Fellows of Harvard College for 1914-15. Cambridge, p. 38.

HEWETT, D. F. and ROVE, O. N., (1930) Occurrences and relations of alabandite. *Economic Geology,* **25,** 36-56.

HILL, C. A., (1981) Higgins Mine Crystal Cave, Bisbee, Arizona. *Cave Research Foundation 1979 Annual Report,* Adobe Press, Albuquerque, New Mexico, 15 p.

HILL, C.A., (1987) Geology of Carlsbad Caverns and Other Caves in the Guadalupe Mountains, New Mexico and Texas. *New Mexico Bureau of Mines and Mineral Resources Bulletin* 117: 1-150.

HILL, C. A. and Forti, P., (1997) *Cave Minerals of the World, second edition.* National Speleological Society, Inc., Huntsville, Alabama, 461 p.

HOVEY, E. G., (1911) Newly discovered cavern in the Copper Queen Mine (at Bisbee, Arizona). *American Museum Journal,* **11,** 304-307.

HUNT, T. S., and DOUGLAS, J., (1887) The Sonora earthquake of May 3, 1887, *American Naturalist,* **22,** 1,104-1,106.

JONES, W. R., HERNON, R. M., and MOORE, S. L., (1967) General Geology of the Santa Rita Quadrangle, Grant County, New Mexico. *U. S. Geological Survey Professional Paper* **555.** Government Printing Office, Washington, D.C., 144 p.

KEMP, J. F., (1900) *Ore Deposits of the United States and Canada, third edition.* The Scientific Publishing Company, New York and London, 481 p.

KUNZ, G. F., (1885) On remarkable copper minerals from Arizona. *Annals of the New York Academy of Science,* **3,** 275-278.

LINCOLN SUNDAY STAR, (1922) *University Museum Geologic Section Gets Rare Specimens.* April 9, p. 16.

LINDGREN, W. and LOUGHLIN, G. F., (1919) Geology and ore deposits of the Tintic Mining District, Utah. *U.S. Geological Survey Professional Paper* **107,** Government Printing Office, Washington, D.C., 282 p.

LOS ANGELES TIMES, (1891) *Huge Chunks of Metal.* March 30, p.4.

LOUGHLIN, G. F., (1926) Guides to Ore in the Leadville district, Colorado. *U.S. Geological Survey Bulletin* **779,** Government Printing Office, Washington, D.C., 38 p.

MASLYN, R. M., (1976) *Late-Mississippian paleokarst in the Aspen, Colorado area*: Colorado Sch. Mines, Golden, Colorado, Geol. Master's Thesis, T-1811, 96p.

MASLYN, R. M., (1977) *Recognition of fossil karst features in the ancient record: A discussion of several common fossil karst forms*, in Veal, H. K., ed., Southern and Central Rockies Exploration Frontiers: Rocky Mtn. Assoc. Geologists Guidebook, p. 311-319.

MCKNIGHT, E. T., and FISCHER, R. P., (1970) Geology and ore deposits of the Picher Field, Oklahoma and Kansas: *U.S. Geological Survey Professional Paper* 588, 165 p.

MILLER, A., (1897) Detail printed on photograph mounting. Grand Lodge of Arizona A. F. and A. M., Globe, by the author, 1 p.

MILLS, C. E., (1956) *History of the Bisbee district.* Unpublished manuscript, Phelps Dodge Corp. files, 16 p.

MILLS, C. E. (1958) *Notations from annual reports (Copper Queen Consolidated Mining Company, Phelps Dodge & Company and Phelps Dodge Corporation) years 1909-1950.* Unpublished, Phelps Dodge Corp. files, 72 p.

MINERAL COLLECTOR, THE (1896a) Anonymously written advertisement for Geo. L. English & Co., **2,** no. 11.

MINERAL COLLECTOR, THE (1896b) Anonymously written advertisement for Roy Hopping, **2,** no. 12, vii.

MINERAL COLLECTOR, THE (1897) Anonymously written advertisement, **3,** v.

MINERAL COLLECTOR, THE (1898) Anonymously written advertisement, **5,** no 4, v.

MINERALOGIST'S MONTHLY, THE (1891) Extracts from William Given's letter to Geo. L. English, 1890. July 1891.

MINERALOGIST'S MONTHLY, THE (1891) Anonymously written advertisement, August 1891, xii.

MOREHOUSE, D. F., (1968) Cave development via the sulfuric acid reaction. *National Speleological Society Bulletin,* **30,** 1-10.

MOORE, T. P., (2006) Richard W. Graeme and the Graeme Family collection of Bisbee minerals and ores. *Mineralogical Record,* **37,** 171-180.

NEW YORK ACADEMY OF SCIENCE, THE, (1889) Transactions of the Regular business meeting, January 7, 1889, **8,** October 1888- June 1889, 45-46.

NOLAN, N. B., (1962) The Eureka Mining District Nevada. *U.S. Geological Survey Professional Paper* **406,** Government Printing Office, Washington, D.C., 78 p.

OSBORNE, R. A. L., (1996) Vadose weathering of sulfides and limestone cave development-evidence from eastern Australia. *Helictite, Journal of Australian Speleological Research,* **34,** no. 1, 5-15.

PALMER, A. N., (1991) Origin and morphology of limestone caves. *Geological Society of America,* **103,** 1-25.

POLYAK, V. J., RASMUSSEN, J. B. T., and ASMEROM, Y., (2004) Prolonged wet period in the southwestern United States through the Younger Dryas. *Geology* **32,** 5-8.

QUINLAN, J. F., (1972) *Karst-Related Mineral Deposits and Possible Criteria for the Recognition of Paleokarst: A review of Preserved Characteristics of Holocene and Older Karst Terranes*: International Geological Congress, 24th, Montreal, 1972, Proceedings, sec. 6, p. 156-168.

RANSOME, F. L., (1904a) The geology and ore deposits of the Bisbee quadrangle, Arizona. *U.S. Geological Survey Professional Paper* **21,** Government Printing Office, Washington, D.C., 168 p.

RANSOME, F. L., (1904b) The geology and copper deposits of Bisbee, Arizona. *A.I.M.E. Transactions,* **34,** 618-642.

RENO EVENING GAZETTE, (1882) September 13.

SATO, M., (1960) Oxidation of sulfide ore bodies. *Economic Geology,* **55,** 928-961.

SCIENCE, (1915) Discovery of the Shattuck Cave. March 12, 387.

SMITH R. M. and MARTELL, A. E., (1976) Critical Stability Constants. **4***: Inorganic Ligands.* Plenum Press, New York.

SMITHSONIAN INSTITUTION, THE, (1891) Annual Report of the Board of Regents for the year ending June 30, 1890, 740.

STEVENS, H. J., (1905) *The Copper Handbook,* **5,** for the year 1904, Horace J. Stevens Houghton, Michigan, 882 p.

STYLES, R. W., (1917) Report on the Geologic Collections. *Annual Report of the Director of the Museum of Comparative Zoology at Harvard Collage to the President and Fellows of Harvard Collage for 1919-20.* Cambridge, p. 32.

TABER, S. and SCHALLER, W. T., (1930) Psittacinite from the Higgins Mine, Bisbee, Arizona. *American Mineralogist,* **15,** 575-579.

TOWER, G. W., Jr., and SMITH, G. O., (1899) *Geology and Mining Industry of the Tintic District, Utah,* in Walcott, C.D., *Nineteenth annual report of the United States Geological Survey to the Secretary of the Interior,* 1897-1998; Part III: U.S. Geological Survey Annual Report, 19, pt. 3, p. 601-767.

TRISCHKA, C., (1926) New cave found in Copper Queen. *The Engineering and Mining Journal,* **121,** 328.

TRISCHKA, C., (1928) The silica outcrops of the Warren mining district, Arizona. *The Engineering and Mining Journal,* **125,** 1045-1050.

TRISCHKA, C., (1932) Ore and the silica outcrops in the Southwest Mine. Unpublished report,   Phelps Dodge Corp. files, 7 p.

TRISCHKA, C. (1938) Bisbee district. *Arizona Bureau of Mines Bulletin* **145,** 32-41 University of Arizona Press, Tucson, Arizona.

TRISCHKA, C., ROVE, O. N. and BARRENGER, D. M. JR., (1929) Boxwork siderite. *Economic Geology,* **24,** 677-686.

WALKER, R. T., (1928) Deposition of Ore in Pre-existing Limestone Caves. *Technical Publication No. 154, American Institute of Mining and Metallurgical Engineers,* 43 p.

WEED, H. W., (1908) *The Copper Mines of the World.* Hill Publishing Company, New York and London, 375 p.

WEED, H. W., (1912) Geology's Relation to Ore Deposits. *Mines and Methods,* **3,** no. 8, 559-560.

WELLS, R. C., (1913) A new occurrence of cuprodescloizite (Bisbee). *American Journal of Science,* 4th series, **36,** 43-47.

WENDT, A. F., (1887) The copper ores of the southwest. *The Engineering and Mining Journal,* **41,** 135-136 & 150-151.

WEYLE, P. K., (1959) The change in solubility of calcium carbonate with temperature and carbon dioxide content. *Geochimica et Cosmochemica Acta,* **17,** no. 3-4, 214-225.

WHITE, W. B., (1981) Reflectance spectra and color in speleothems. *The NSS Bulletin,* **45,** 20-26.

WHITE, W. B., (1997) Color of Speleothems, in Hill, C. A. & Forti, P. *Cave Minerals of the World,* second edition. National Speleological Society, Inc., Huntsville, Alabama, 463 p.

WHITE, W. B., (2007) Cave sediments and paleoclimate. *Journal of Cave and Karst Studies,* **69,** no.1, 76-93.

WILSON, P. D., (1914) A cavern in the Shattuck Mine. *The Engineering and Mining Journal,* **97,** 743-744.

WISSER, E. H., (1927) Oxidation subsidence at Bisbee, Arizona. *Economic Geology,* **22,** 761-790.

COPPER CZAR
PUBLISHING

www.ingramcontent.com/pod-product-compliance
Lightning Source LLC
Chambersburg PA
CBHW051600190326

41458CB00029B/6493